FARM ANIMALS
in the
MAKING OF AMERICA

By
Paul C. Johnson

Published and Manufactured
in the
United States of America
by the
Wallace Homestead Book Company
P.O. BOX BI
Des Moines, Iowa 50304

Library of Congress Catalog Card Number 75-366-36
ISBN: 0 87069 135X

COVER PHOTOS: Front cover is an old lithograph of a team of Arden Horses, used as a frontispiece in the 1856 report of the Commissioner of Patents, forerunner of the U. S. Department of Agriculture. Back cover is one of the famous hen and rooster paintings distributed by the Poultry Tribune, this one showing a pair of Partridge Wyandottes.

CONTENTS

DEDICATION

THIS BOOK IS DEDICATED to my brother Maurice and his wife Lil who have operated the home farm in Dakota County, Minnesota, and who have worked long and hard to nurture and improve the ancient art of livestock husbandry. Maurice is a born stockman who has loved his animals and given unstintingly of his time and skill to insure their well-doing. He has made special contributions in developing the breeds of Shropshire sheep, Percheron horses and Holstein cattle. Lil exercised great patience and understanding whenever the going was hard and when the farm animals got more attention than she did. Together they raised a fine family, and served their church, their farm co-operatives, and their community.

FOREWORD

This book is made up largely of illustrations and text material from old farm magazines, amplified by comment and reminiscences of the author who, while serving as editor and editorial director of *Prairie Farmer* from 1947 to 1970, had a golden opportunity to delve into the glorious past of American Agriculture. *Prairie Farmer,* founded in 1841 and still going strong, has almost complete files going back to the date of its origin. But there were many others, such as *American Agriculturist,* still being published, *Orange Judd Farmer* and *The Cultivator,* pioneer papers which have faded into history. I confess it was the magnificent hand done engravings in the old publications that impelled me to lift them out of the old files and make them available to today's readers. The plates were made out of wood, copper and steel for use with hand set type in printing the publications of the nineteenth century. They were painstakingly and beautifully wrought by artisans who worked mostly in the back rooms of magazines and newspapers. Considering the crude printing presses and primitive paper, the illustrations have held up remarkably well for 100 years or more. In addition there are some very good stone lithographs that were printed separately and used as frontispieces and inserts in publications of a century ago.

The engravings that were reproduced in the farm papers before the Civil War were necessarily simple

BELOW is a steel engraving taken from a book entitled "Facts for Farmers," written by Solon Robinson and published by John and Ward, New York, in 1864. It was full of information covering the full spectrum of agriculture of the time and undoubtedly was one of the books sold from house to house by peddlers, along with Bibles and doctor books. The engraving is reminiscent of the Currier and Ives engravings of rural life which were distributed in great numbers a short time later.

5

and rather crudely done. There was steady improvement, however, until the 1880s when the illustrations in such papers as *Prairie Farmer, Orange Judd Farmer, American Agriculturist,* and *Breeder's Gazette* were truly beautiful. They put a kind of bloom on the period which I will venture to call The Golden Age of American livestock breeding.

I hope the reader will celebrate America's Bicentennial by enjoying the superb illustrations, and by absorbing some of the spirit of the great livestock breeders. For them farm animals were not just meat and hide and pounds of production. They were subjects of the art of animal husbandry when it was much more of an art than a business.

The love of good livestock, the creation of types and blood lines more by eye and instinct than by any exact science of genetics, judicious use of crosses and line breeding,—all these blended to produce not only remarkable animals but a culture and a camaraderie that has all but disappeared today.

Some would dispute the above statement, but there is no room to argue the point here.

About the middle of the 1890s, photoengraving, the mechanical process of transferring photographic images to printing plates, was perfected to the point where it was available to farm magazines. The quality of illustrations took a sharp nose dive. Perhaps the original pictures were poor, the engraving sloppy, or it took time to learn to use the screened halftone plates properly on the printing presses of the period. Anyway, I have used very few pictures from the period after 1895 because the original photographs are unavailable and the printed pictures were too poor to reproduce cleanly.

Livestock photography as an art also took time to develop. The engravers had their own stylized approach to their subjects. The result was often grotesque. They exaggerated blockiness, shortness of legs, and other features to create a beast that was often more imaginary than real. But they were very good at it, as you can see from the illustrations in this volume. Later the livestock photographers were to try to achieve the same effect by standing their animals deep in straw and taking great pains with their angles.

Very few of the engravers and lithographers of livestock subjects are known by name. They probably worked cheaply, because most of the farm publications were not affluent. They must have worked fast because often they had to come up with illustrations in a hurry as news events broke. Farm papers were then, as now, more news magazines than journals of scholarly information. While the best engravings of the nineteenth century undoubtedly turned up in such general publications as *Harper's Weekly, Scientific American,* and the various women's magazines, the farm papers did a very respectable job. They recorded an exciting period in the history of American Agriculture with dash and appreciation.

While old bound copies of farm papers from the last century are disappearing fast, there was once a rather good supply. The great danger is that libraries, pressed for space and obsessed with the idea of getting everything on microfilm, may destroy many of the precious old volumes.

An odd but significant quirk in agricultural history is responsible for the availability of bound farm papers of a hundred or more years ago. In the days before the agricultural experiment stations and extension services, farmers did their own experimenting with both livestock and crops. They also invented their own farm machines. They were eager to exchange ideas with those of like interests. Much of the editorial content of farm papers was written by farmers. *Prairie Farmer's* slogan for the first half-century was,

"Farmers, write for your paper!"

There was a voracious appetite for reading of any kind, but reading matter was in short supply. Editors and publishers of the day found that they could augment their income by printing extra copies of their papers, binding them (often in leather) at the end of the publishing year, and selling these to readers. The most common price was two dollars a volume, which was a really remarkable price when one considers that many hard bound books could be had for fifty cents.

It is not easy to comprehend today that people of a century ago would eagerly buy and read year-old publications. Historians have reason to be grateful.

When *Prairie Farmer* lost most of its files in the great Chicago fire of 1871, it was comparatively easy to restore those files by appealing to readers who owned bound volumes from past years.

I feel strongly that historically minded readers of today should have access to this gold mine of fascinating reading matter and superb illustrations. I have tried to pass on in this book some of the best of the engravings and most provocative of the ideas. This is a good time to study our past and understand better whence we came.

OPPOSITE PAGE

BACK IN THE 1880s Prairie Farmer ran a series of charming front page engravings celebrating farm activity on the farm in a particular month. Farm magazines of the past carried a great variety of material and ran to the "literary" more than they do today. The whole spectrum of the interests of the farm family was covered in one way or another.

THE PRAIRIE FARMER

A Weekly Journal for

THE FARM, ORCHARD AND FIRESIDE.

ESTABLISHED IN 1841.

ORANGE JUDD,
Editor and Business Manager.

CHICAGO, ILL., SATURDAY, MAY 8, 1886.
OFFICE, 150 Monroe Street.

PRICE, 5 CENTS—$1.50 PER YEAR. } No. 19.
IN ADVANCE, DURING REST OF 1886. } Vol. 58.

MAY.

I feel a newer life in every gale,
  The winds that fan the flowers,
And with their welcome breathing fill the sail,
  Tell of serener hours—
Of hours that glide unfelt away
Beneath the sky of May.

The spirit of the gentle south wind calls
  From his blue throne of air,
And where his whispering voice in music falls
  Beauty is budding there;

The bright flowers of the valley break
  Their slumber and awake.

The waving verdure rolls along the plain,
  And the wide forest weaves,
To welcome back its tuneful guests again,
  A canopy of leaves.
And from its verdant shadow floats
A gush of trembling notes.

Fairer and brighter spreads the reign of May;
  The tresses of the woods
With the light dallying of the west wind play,
  And the full brimming floods,

As gladly to their goal they run,
Hail the returning sun.—*James G. Percival.*

THE MAYPOLE.

It is the choice time of the year,
  For the violets now appear;
Now the rose receives its birth,
And lovely flowers deck the earth.
Then to the Maypole come away
For it is now a holiday.
    —*From an Old English Song.*

HYMN ON A MAY MORNING.

Now the bright morning-star, Day's harbinger,
Comes dancing from the east, and leads with her
The flowery May, who from her green lap throws
The yellow cowslip and the pale primrose.
  Hail, bounteous May, that doth inspire
  Mirth, and youth, and warm desire!
  Woods and groves are of thy dressing;
  Hill and dale doth boast thy blessing.
Thus we salute thee with an early song
And welcome thee and wish thee long.—*Milton.*

INTRODUCTION

What I choose to call the Golden Age of livestock breeding was already beginning to wane when I was born on a Minnesota farm in 1904. Nevertheless, I did get a wealth of first hand exposure to the people, the talk, and the camaraderie of the show ring. The years just before and during World War I were an exciting time on the farm. Income was relatively good, the county and state fairs were in their heyday, and the do-it-yourself improvement activity of the 1880s was giving way to the better organized programs of the agricultural colleges, experiment stations, and U. S. Department of Agriculture.

We didn't know it at the time, but the ground was being laid for the highly mechanized, production oriented farming we have today. The farmer as an individual dreamer and experimenter and—may I say—*lover* of his chosen field of husbandry, was to become less involved in the creation of his seed stock and the shaping of his product. Specialization, or the restriction of activity on any one farm to fewer crops and fewer kinds of livestock, was coming up on the horizon, though yet some years away. The farm which had about everything in the way of crops, fruits, farm animals and poultry would soon be a thing of the past.

While the hybridization of both crops and livestock was still some years away when farmers experienced the World War I boom in purebred livestock prices, change was in the air. I recall that my father, with his sixth grade education but an unusually open mind about farm improvement, had a successful and profitable cross breeding business going in hogs as early as 1912. The Poland China hog at the time was a big, lardy animal which made fairly good gains but did not have an especially good reputation for mothering ability and size of litter. Chester Whites did much better in this department. (There may have been some Yorkshire blood in the Whites we knew.) Father conceived the idea of crossing a Poland boar on White gilts. Thus he developed a pork raising business that literally paid off the mortgage on the farm.

OPPOSITE PAGE

THIS FRONT PAGE represents The Breeder's Gazette in its full glory during the Golden Age of American livestock breeding. The ornate logo was characteristic of all agricultural magazines of the time. Breeder's Gazette had the distinction of serving the full spectrum of the livestock industry in contrast to the more restricted "breed magazines" that have served individual breeds from the time of the organization of the breed associations down to the present. The general farm papers also had to restrict somewhat their livestock coverage and enthusiasm because they covered crops as well as livestock. They did, in fact, serve the interests of the whole farm family.

BELOW is an engraving from the Chicago Horseman of 1885, revealing the stylized approach to illustration of livestock which was common at the time. The best artist-engravers were superb in their capture of the flowing lines of the running horse— indeed they put life into the farm animals that they drew. In contrast there were some, notably Currier and Ives' Fannie Palmer, who simply could not draw a good horse, even if her depictions of the pioneer American farm are among the most charming of all.

THE BREEDER'S GAZETTE

A WEEKLY JOURNAL OF LIVE STOCK HUSBANDRY.

Vol. XVIII. CHICAGO, ILL., NOVEMBER 26, 1890. No. 22.—470.

Rodin (3434)

THE BRADLEY "HANDY" WAGON.

(PATENTED 1885.)

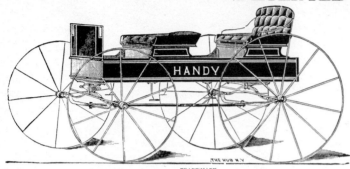

TRADE MARK.
THE BRADLEY "HANDY" WAGON.

THE BRADLEY
"Handy" Surrey

is hung on the same style gear as our "Handy" Wagon, and by doing this we produce a four-passenger pleasure vehicle much lighter in weight than any other of the same capacity, and consequently easily handled by a single horse. We produce this Surrey in two grades, bringing our prices within the means of every horse owner. Roomy, easy to get into and out of, convenient, durable, well-finished, and not expensive. Send for Catalogue—free.

As it is not possible that every dealer in vehicles in this country will carry a stock of our "Handy" Wagons, etc., we would suggest that you send to us direct for our Catalogue, which will be promptly mailed, and in case your local dealer is unable or unwilling to quote prices, we will name a price at which we will deliver any vehicle of our make to any R. R. station east of the Mississippi River.

THE BRADLEY
"Handy" Wagon

has met with unprecedented success since its introduction, and is, beyond a question, the most popular vehicle of its class in the market. We furnish this wagon with one or two seats (both movable), shafts or pole, and with or without canopy. The special feature of this wagon is in the style of spring used, and its connection with the axle. The axles are cranked down with the springs directly over them and connected to them near the wheels, giving the greatest strength with the least possible weight, and allows the body to hang lower than in any other spring vehicle in use. In its construction we use nothing but **Solid Steel Axles**, the finest **Oil Tempered and Tested Steel Springs** and **Wheels** that we guarantee in the strongest manner. A thoroughly well-made wagon. **Simple, Light, Convenient** and **Low Priced.** Send for Catalogue—free.

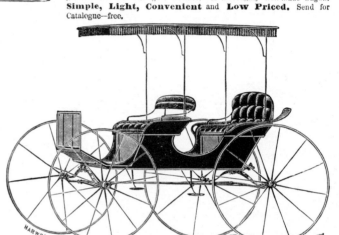

THE BRADLEY "HANDY" SURREY.

THE BRADLEY
"HANDY" BUCKBOARD

is another style of our "Handy" vehicles, and is especially designed for use in mountainous sections, or over roads where lightness of weight is a great consideration. As this is the simplest form in which it is possible to construct any four-wheeled vehicle, we are able to name exceedingly low prices on our "Handy" Buckboard, and at the same time furnish a thoroughly durable and satisfactory job in every respect. Send for Catalogue—free.

While we prefer to furnish our vehicles through regular dealers in carriages throughout the country, it is always a good idea for the consumer to have a copy of our catologue, which we mail free upon application, as in it we are able to accurately illustrate and describe our full line of work. We have sold thousands of our vehicles direct to consumers with perfect satisfaction to everyone concerned.

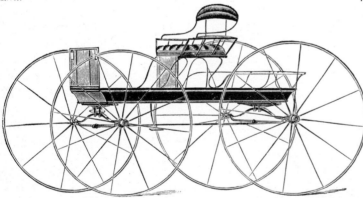

THE BRADLEY "HANDY" BUCKBOARD

The Bradley Two-Wheeler,

which we produce in several styles, has gained the reputation of being the **best vehicle of its class in the world.** This is a strong statement, but we are willing to place them in the hands of our customers on that basis. We guarantee them to be **absolutely free** from what is known as the disagreeable **horse motion**, or no sale. As dealers in vehicles do not carry this class of work in stock as a rule, a large proportion of our sales are made direct to consumers, and we guarantee satisfaction in every case. Send for Catalogue—free.

The Bradley Road Cart we are not able to illustrate here from lack of space. We have recently greatly reduced our prices on this line of vehicles, and every farmer who has a colt to break, every trainer of trotters, and every owner of a horse, whether speedy or not, should have one. Send for Catalogue—free.

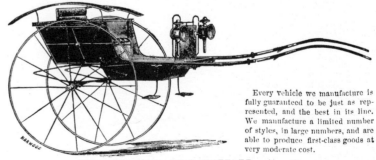

Every vehicle we manufacture is fully guaranteed to be just as represented, and the best in its line. We manufacture a limited number of styles, in large numbers, and are able to produce first-class goods at very moderate cost.

THE BRADLEY TWO-WHEELER.—No. 6.

10

Along about 1916 the older boys went off to Agricultural School and came home deeply imbued with the purebred livestock spirit. They announced to dad that they would not tolerate those spotted scrubs on the place. So father shifted to purebreds and the hog business went into decline—at least that is his version of the story.

At that time the livestock professors at the universities were almost to a man advocates of purebreds. A few were beginning to experiment along other lines. In 1926, Professor L. M. Winters of the University of Minnesota, who was on the faculty where my brothers drank deeply at the fountains of the purebred tradition, published a bulletin entitled *"Calico Hogs,"* which advocated cross breeding for greater vigor and better gains. Winters then started the research that later produced his Minnesota No. 1, combining the old tools of cross breeding and inbreeding to produce an animal tailored to the growing market for lean pork.

Indeed, these *were* old tools. They had been used freely by the early creators of livestock breeds, both in Europe and America. Virtually all our breeds trace their origins to crosses of one kind or another. Often these were the accidents of migration of people and animals from one geographical area to another. But early in the history of animal husbandry there were deliberate attempts to achieve desired results by crossing. Horses were bred for war as well as work and sports. Cattle were bred to large size for both draft and meat purposes; occasionally, as in Scotland, for hardiness in an unfriendly climate. Hogs were bred most often for lard. The breeds, and the lines within breeds took shape under the skillful and—I say again—loving hands of the breeders.

In early years—say the first half of the 1800s and earlier—there was greater tolerance of out-crossing, that is finding desirable characteristics wherever they could be found.

The fierce jealousy between advocates of specific breeds and the battles to keep breeds pure were later developments. The growth of restrictive practices to protect breed purity may have been due to the protective instincts of breed associations which were the first organized policemen on the scene.

By the early 1900s it had become a kind of religion to keep breeds pure. While the desirability of an occasional outcross was recognized and very reluctantly tolerated, most of the search for better genes outside the breed was conducted "behind the barn," so to speak, with some falsification of papers.

When my brothers and I were in our teens on the farm, we developed quite a good flock of grade Shropshire sheep. Some animals were so good in quality that they consistently beat the purebreds at our county fair, which maintained for a time mixed classes to encourage general livestock improvement. A purebred breeder living nearby got into the habit of coming over and paying premium prices for some of our best grade ewes. We put two and two together and concluded he was supplying live animals to fit pedigrees that had been orphaned by one mishap or another. The practice undoubtedly improved his purebred flock as to quality, but we boys would never have thought of making such a substitution in our own flock. Soon we got rid of the grades and were the proud possessors of only purebreds. We had "arrived."

Sometimes the fierce protectiveness to keep breeds pure was carried to a point where a whole breed picked up impractical characteristics and almost ran itself in the ground. Fortunately, there developed in nearly every breed "strains" or "lines" based on geography or the genius of a particular breeder. Crossing these strains served to perk up the breed and introduce a small dose of that quality known to hybridizers as heterosis—the boost that comes from crossing two genetic bundles some distance apart.

The breeders worked hard at improving their stock. Somehow, they managed to make real progress in spite of some of the ill effects of breed patriotism. Their diligence in picking practical characteristics such as vigor, rapid growth, and acceptance by the consumer fitted the purebreds for the important role they were to play in furnishing the foundation stock for the cross breeders and the hybridizers. In both the plant and livestock worlds, the search for good "protoplasm" is endless. The pure breeds still play an important part. We keep going back to them. There are still new chapters to be written in the search for better farm animals, and I feel sure the purebreds will have an important place.

It is not my purpose here to furnish a definitive history of the breeds of livestock, nor of the great breeders. There are some excellent books available now and more need to be written. My purpose is rather on the occasion of the American Bicentennial to give more Americans a glimpse into a facet of American history that is much neglected by the formal historians.

Horses were almost as important as their human masters in opening up the country, felling the forests, breaking the prairies, and transporting people and goods. Cattle not only served as draft animals, but they supplied milk that turned stone-ground meal into nutritious mush. They became the chief converters of the oceans of western grass into meat for use at home and barrels of salt beef that were among the first exports back East and to Europe. I can extend this eulogy to the hog and the sheep, as well as the poultry. All traveled with the covered wagon.

While pioneers had from necessity a no-nonsense attitude toward the farm animals which fed them and earned for them badly needed "cash money," there was also evident in nearly every facet of pioneer farm life a kind of personal affinity between man and his livestock. This undoubtedly goes all the way back to the prehistoric period when man found both profit and companionship in domesticating the animals around him.

It has been said that life had to be a lot more interesting in the old days because there was always present the interplay of people and the animals they worked with. Anybody who has lived with livestock knows that they are individuals—characters if you please—with ways and wills of their own. Traditionally, we have attributed such "character" mostly to horses and dogs, occasionally to "faithful oxen." But that is doing a

BELOW

THE CARRIAGE HOUSE was one of the familiar "out back" buildings in every town or city at the turn of the century. Today many are still in existence but they have been remodeled into studios and apartments much coveted by young couples looking for a place to rent. They were really glorified barns, housing the buggies as well as the horses that any well-to-do family needed for transportation and for prestige. As often as not there was a manure pile which furnished a supply of fertility much appreciated by the gardener.

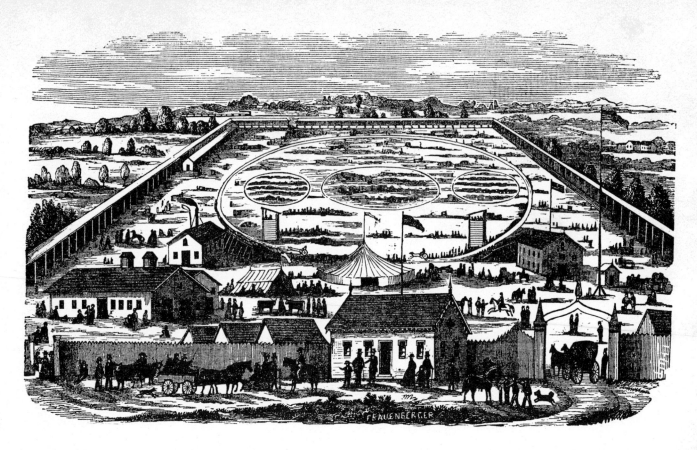

great injustice to other species of domesticated livestock. Farm animals were not just "peas in a pod." They differed individually.

There was the nuisance horse who could figure out almost any latch on a door or gate and let himself and companions out of confinement. There was the bull who always thought the heifers on the other side of the fence had more sex appeal, expecially if they were a different color. There are a thousand stories of farm animals that have done amusing and unusual things, even influenced the course of history.

When I was a farm editor I used to run a contest periodically and offer a prize for the best essay on "The Smartest Animal I Ever Knew." Farm folk responded with great humor and not a little literary talent. Every animal had its day, not least the lowly hog which is reputed to be the smartest of all. Deliver me from starting an argument here!

The "personality" characteristics of farm animals probably had a great deal to do with the love which good stockmen developed for the animals they bred and improved and often pampered. Mostly stockmen did fairly well in the economics of farming, but many went broke. There can be no doubt that this resulted in many cases from their determination that animals should have the best available whether the breeder could afford it or not.

As a farm editor during the period when genetics became an advanced science, and we succeeded in blurring the breeds and homogenizing livestock production in the interests of rate of gain and profitability of product, I expect I had a hand in the "destruction" of the old livestock tradition. If so this book is my penance. I have known many livestock people and found them to be rich in human traits that lend interest to life and excitement to history.

A good stockman was a breed apart. Either you were one or you weren't. I hesitate to put myself in

ABOVE

COUNTY AND STATE FAIRS were a most important social and educational event in pioneer days. Farmers organized these events almost before the land was cleared and the Indians were persuaded to move farther west. This rare woodcut depicts the New York State Fair at Rochester, way back in 1862. It was taken from the Country Gentleman of that date. Livestock of every category were the heart of the fair. Nearly every fair had its race track to attract those with sporting blood, which included practically everybody.

the stockman category. I don't have what it takes.

When my older brother and I were in our teens, taking turns going away to school while developing our own livestock enterprises as well as father's, it was routine on winter evenings to milk the dairy herd, give all the stock their evening feed, "batten down the hatches" against the winter's cold, haul the milk cans to the milk house, and call it a day. I would hurry through these chores and proceed to the house to bury my nose in the *American Boy* or other reading matter that was at hand. Instead of coming at once to the house, my brother would linger to observe the animals one by one, spread a little extra straw to insure greater comfort and cleanliness, go from barn to barn with the kerosene lantern to see that all classes of livestock were comfortable, watered and fed. When he finally showed up in the house, he would study pedigrees and pore over livestock journals. He was a stockman. I wasn't.

This book will deal only with the principal livestock species that had the greatest economic value on the American farm, and especially the Midwest with which I am most familiar. They are horses, cattle, hogs, sheep, and poultry. I must apologize for leaving some out. The goat, for instance, deserves consideration because it is a very versatile and intensely interesting animal, of real economic importance in many parts of the world. We just didn't raise many goats down our way. Other readers will lament the absence of a chapter on dogs. They too helped build the country as companion and servant of man. Even cats were work stock. They were expected to keep the farmstead clear of mice and rats (even gophers) in exchange for a squirt of milk at milking time and a place to call their home.

It is fitting here to acknowledge our debt to the stockmen and livestock fanciers of Europe. Before America was colonized they had established a notable livestock tradition and set the stage for what American breeders were to do later.

Most of our livestock breeds originated in Britain, France, and the Low Countries. It was only in this century that we embarked on a worldwide search for genetic material to enhance today's livestock industry, adding disease resistance, heat tolerance, hardiness, size, and other significant traits. You will be interested, however, to learn that breeds such as Charolaise and Simmenthal, which have become popular today, have been around for a long time. The early breeders somehow passed them by.

In America they were looking for grain-efficient animals to use up our vast quantities of corn, oats, barley, and oil seeds. The future search may be for grass- and roughage-efficient animals, even as we shifted gradually and painfully in dairying from the butterfat producers to the protein producers to better suit our sedentary population.

In selecting materials for this volume I have digressed to give the reader a number of things that, I believe, have historical significance as well as charm and beauty. Note, for instance, the elaborate engravings that graced the mastheads of the early farm magazines. Would that such art could be restored in our modern publications! I have noted earlier that the illustrations used here are largely from that period when hand done engravings were the hallmark of all the best publications. I want more Americans to know that the Golden Age of livestock was also a Golden Age of livestock journalism. There were many more farm papers then than there are now. They had the same goals, a better agriculture and a better life for the farm family. But they were written more by amateurs than by pros, with interesting results.

THE RACINE WISCONSIN AGRICULTURIST.

THE BEST 50¢ PAPER IN THE WORLD.

VOL. XIII.—No. 1.   RACINE, WISCONSIN, JANUARY, 1889.   50 CENTS A YEAR.

New Years Day in the Country.

Warming Water For Stock.

The former theory of warming water for stock is a theory no more, and the question, Does it pay? is resolved into the query: What and where is the best tank heater? This is as it should be, and the Agriculturist congratulates its readers upon their humanity and intelligence. Time and practice have shown the advantages of warm water; and ice water, shivering animals, shrunken milk supply and double feed are things of the past in a progressive dairy.

Though some few say it is useless, the majority of stockmen, the thinking, money-making portion, claim gains in less food required to sustain the temperature of the body, comfort of the stock (which no sensible man can deny) and in more and richer milk.

If the question were only one of the humanity of the practice we should urge it, but as it is, affecting both the feelings and the purse, we must urgently call attention to it.

The accompanying engraving illustrates, (and quite truthfully too) the difference in appearance of stock after a winter of ice water on one hand and warm water on the other.

In purchasing a heater, as in all other things, there is more than the first cost to be considered. The furnace, to be durable, must be substantially made and capable of withstanding both fire and water. It should be so constructed as to utilize all the heat possible in warming the water, and as the water is frequently low in the tank, the heat should be at or near the bottom of the heater.

When at Rockford, Ill., a few days since, being desirous of obtaining further information of heating and heaters, a representative of the Agriculturist called upon Messrs. Shoudy & Miller, of that city, the manufacturers of The Lightning Base Burning Tank Heater and Feed Cooker, with which many of our readers are doubtless acquainted; but for the benefit of those who are not, we present a few points of excellence in this successful heater, which the manufacturers inform us has occupied their attention for the past several years and has only recently been perfected.

The heater, as the name implies, is a base-burning furnace which may be placed in a water tank to warm water or in a large tub for cooking grain. The flame from the fire box sweeps across the bottom of the heater underneath the ash box, heating or cooking with great economy of fuel and creating a circulation of water in the tank without employing a coil of pipe.

The great objection to nearly all coal-burning heaters is, the fire must be extinguished to get the ashes out, consuming half an hour's time before the fire is again rekindled. This objection has been overcome in The Lightning Heater by means of a connected sectional ash-pan, which removes the ashes without disturbing the fire.

These heaters are built of heavy sheet metal and lined with castings to protect the outer casing from the direct heat of the fire.

All machines have their good points, but seldom do we see one combining all so completely as does this heater, and Messrs. Shoudy & Miller are to be congratulated upon the perfection to which they have brought their furnace, and upon the perfect satisfaction they are giving to users. Those in need of such an article as this should write for a descriptive circular which will be mailed to any address on application.

Failure to have plenty of exercise, too much draught, sudden changes in the weather, all have more or less influence with poultry in maintaining good health, and when disease makes its appearance it is a good plan to investigate, and as far as possible ascertain and remove the cause. In many localities there is no question but that ducks can be made more profitable than chickens. They will with good treatment lay more eggs, make more rapid growth and sell in a market at better prices while they are far less liable to disease.

15

THE PRAIRIE FARMER

A WEEKLY JOURNAL OF THE FARM ORCHARD AND FIRESIDE

TERMS:
$2.00 per Year.

"FARMERS, WRITE FOR YOUR PAPER"

OFFICE:
112 Monroe Street.

ESTABLISHED IN 1841.   CHICAGO, SATURDAY, APRIL 8, 1871.   ENTIRE SERIES, VOL. 42—No 14.

DUKE DE CHARTRES.   DENMARK.   MASTODON.   INDEPENDENCE.   VIDAL.

JAMES A. PERRY'S GROUP OF IMPORTED DRAFT STALLIONS, IMPORTED BY HIM AUGUST 17th, 1870, FROM NORMANDY, FRANCE, and ENGLAND, to Clifton, Iroquois Co., Ills.

DESCRIPTION OF ANIMALS SHOWN IN ILLUSTRATION.

DUKE DE CHARTRES is five years old, a dark iron grey color, sixteen and one-half hands high, and weighs nineteen hundred pounds. He has fine style and action. He has a beautiful head and neck, deep shoulders, broad across the breast, hips, and stifle, heavy bone and muscle, and is in every respect a model draft horse.

DENMARK is four years old, of a light roan color, with dark legs, mane and tail, sixteen and one-half hands high, and weighs seventeen hundred pounds. He is very high on the shoulder, heavy breast, heavy limbs, short back, with a large loin, broad across the hip and stifle. The above cut is a perfect picture of this horse as seen by his many admirers. He is also lively and active in his movements.

MASTODON was purchased in Essex County, England; is a pure bred English draft horse, three years old past, is a bright bay color; seventeen hands high, and weighed when he landed in New York nineteen hundred and forty pounds. He has a beautiful head, with a fine arched neck, heavy bone and muscle, finely developed limbs, with good style and action. Girth eight feet and three inches, and measures more around the stifle than any other horse in the United States. Present weight twenty two hundred pounds, and he is probably the largest stallion in this country. He has a symmetry of form that fits him for an artist's model.

INDEPENDENCE is five years old, iron grey color, sixteen hands high and weighs seventeen hundred pounds. He is very high on the shoulder, heavy brest, heavy limbs, short back, with a large loin, broad across the hip and stifle. The above cut is a perfect picture of this horse as seen by his many admirers. He is also lively and active in his movements.

VIDAL is five years old, a beautiful dapple grey color, seventeen hands high, and weighs seventeen hundred and fifty pounds. He has a very fine head and neck, symmetrical limbs, broad chest, deep through the shoulders, short back, close ribed, a full heavy loin, well coupled, long hip, broad across the stifle, of fine carriage and action, and is in every respect a horse that will suit the breeder of draft horses.

These horses are all of a quiet and gentle disposition, easily managed, and well broken to the harness, and were all selected with great care. They were all purchased and shipped at one time from Havre, France, on steamer Denmark, Capt. Forbes, Commander, being the last shipment of horses made after the breaking out of the Franco-Prussian war.

MICHIGAN AGRICULTURAL COLLEGE.

In a recent lecture, before the faculty and students, Professor A. J. Cook, spoke of the refining power of well tended gardens and lawns and trees. The great outlay of labor to beautify the college grounds, is more than paying its way by means of its esthetic influence upon the students and visitors. We should study to make ourselves, our houses and surroundings agreeable. Contrast the effect upon young people of a well managed home, where there are beautiful trees and gardens, with a bleak yard abounding in weeds, briars, dead grass, mustard and a few poppies and sunflowers. Birds are an additional attraction; besides their monied worth as insect eaters, their study and care beget in children, habits of mercy, kindness and purity. Birds in our yards at the college, are very abundant, for they enjoy undisturbed freedom and become so tame that it is often a matter of surprise to visitors. The students feel an ownership and are interested in their protection. The docility of cattle, sheep and other domestic animals, belonging to the college, has often been admired.

A large part of the lecture consisted in a plea for the birds. Even the robin, which strips our currant bushes, pilfers our strawberries and takes nearly all of our cherries, for three months of the year, feeds itself and young upon insects and worms exclusively. Numerous authorities were cited and facts given to prove that we should protect the cedar-bird, black-bird, jay, crow, sparrow and all other birds without exception. We have no birds which do not more than pay for the fruit they eat, by eating worms and slugs and insects.

It may be added that some members of the faculty differ with Professor Cook, on the wholesale approval of all birds.

The members of the chemistry class are jubilant over the passage of the bill by the legislature, granting money for a new laboratory. It is very much needed.

The present freshman class numbers over eighty members, and still they come. They are unusually well prepared. Over seventy-five per cent. of them are sons of farmers or mechanics. The largest freshman class of any former year, we are told, consisted of thirty six students. On account of the peculiarity of the course of study, the faculty—some of them at any rate—do not wish to see over two hundred students at a time in the college. Should the number become very large, they fear the labor system would not be as successful as at present.

Mr. Richard Haigh, Jr., a former graduate of the college, has been appointed Secretary pro tempore, with full power to act in the place of the late Sanford Howard, Secretary of the Michigan State Board of Agriculture. For several years, Mr. Haigh has filled the position of assistant secretary with eminent success.

W. J. B.

TREE PLANTING IN NEBRASKA.—The Blair Register says: Farmers are already preparing for tree planting. One day last week about 16,000 young trees passed through town on their way to farmers farther west in this county. One company of Swedes, from Bell Creek, had 12,000 cottonwood trees from the river bars, while the balance of the lot was in the hands of a farmer nearer by, and consisted of cottonwood, maple and box elder.

THE YEAR 1871 was an eventful one for farm papers which by that time pretty well dotted the country east of the Missouri-Mississippi Rivers. It was especially eventful for Prairie Farmer in Chicago. In April the paper took note of the importation of a group of exceptionally fine stallions from England, France, and Normandy by publishing this front page with its elaborate engraving. In August the editors virtually hosted a nationwide "Convention of Friends of Agricultural Education," namely professors and university leaders from all over the country. Prairie Farmer editors provided shorthand coverage of the meeting and volunteered to publish the complete Proceedings in its pages. This meeting is credited by many to be the origin of the National Association of State Universities and Land-Grant Colleges, which was actually launched as a formal organization in 1887.

No sooner had the Proceedings been published [spread over several issues so other news would not be completely crowded out], when tragedy struck. On October 8-9 Chicago nearly burned to the ground in one of the great city fires of all time. With it went the Prairie Farmer building and equipment, including the woodcut from which this front page was printed. The paper literally rose from the ashes and published within a week its next issue, using borrowed type and borrowed presses from neighboring towns.

HORSES

The horses that had such an important part in the making of America were of ancient and honorable lineage. The horse has been in on practically every important event in recorded history. If you include as a part of the family the lowly ass, which had served mankind for centuries before it was called upon to carry the Mother of Jesus to Bethlehem, and the mule which is a hybrid of the horse and the ass, you can safely say that these beasts had almost as much to do with the course of history as man himself.

We have rather complete fossil remains of the horse that date from the pre-glacial period, and fragments that go back much further. The horse is believed to have originated in Asia and migrated to the Americas by the land bridge at the North, more or less simultaneously with the migration of man over the same route. But something happened to wipe out completely all vestiges of the horse family in both North and South America. There is speculation that some kind of plague was responsible.

RETURN TO AMERICA

Horses did not return until the Spanish explorers and conquerors took over parts of North and South America after Columbus. They brought their war horses with them. Some went wild and later turned up in history as mustangs and Indian ponies, somewhat smaller than their Spanish antecedents, but destined to resume their warlike role as the Red Man battled to save his land from the encroaching white colonists.

The horse family, the *Equidae*, which includes asses and zebras along with horses started its partnership with man well before recorded history. Of the three branches only the zebra refused domestication, and does so even today. Hybridization takes place between members of this family, but the offspring are usually sterile. Only the horse-ass hybrid, known as the mule, has played an important part in the history of man.

There is record of asses being beasts of burden in caravans of the East as early as 3000 B. C. Drawings from caves and tombs show members of the horse family as among the first domestic animals. The earliest records of history find the ass and mule important as pack and draft animals, and the horse in the more exalted role as fighter. This was true both in Asia and Europe. In fact, there developed a phenomenon known as the "Horse Civilizations."

As far as I know, nobody has done a really adequate job describing the part played by the horse in these civilizations that moved on horseback. From the tidbits that have come to us through legend and literature, we know that each tribe or nation bred its own horses, that some horses were large and some small. The terrible Huns who came out of Asia and rolled over

A CAPACIOUS OCTAGON BARN.

The barn, of which Figure 1 gives a perspective view, is forty-four feet across from side to side, with a wing twenty-eight by fifteen feet. The basement, Figure 2, is of stone, laid in lime mortar, but in situations where gravel and cement are abundant it may be built of concrete. Passage ways, A, are provided with tramways upon which a feed-car may be run, affording ready communication between the feed-room, B, and the various stalls, of which those marked C are for horses and F for cattle. At the intersection of the tramways is a turntable, I, by means of which the feed-car may be run upon either lateral track. The spaces marked H may be used for box stalls or pens for pigs, sheep or calves. The feed-room, B, is provided with the boiler, K, a chest, L, for cooking and steaming feed, and the water-tank, M, which may be supplied by rain water from the roof or a pump operated by a windmill. In the rear of the horse stalls at the left of the main entrance is a harness closet, D, well lighted and roomy. Adjoining are the calf-pens marked E, and at the ends of the further range of cattle stalls are two others. The inner and outer doors are all indicated by a letter J. The basement of the wing, N, is devoted to corn-cribs and the storage of wagons and implements. Adjoining it and beneath the inclined driveway is a root cellar, O. One section of the roof is extended downward to form an open shed, P.

The main floor, Figure 3, is devoted principally to the storage of hay and grain. The granary, A, is furnished with bins for

extends across one side of the larger room, filled from the outside room. The drive-way and bridge are respectively marked J and I.

The barn was designed by Sheldon F. Smith, York Co., Pa., who writes us as follows concerning the general plan: "The live-stock of whatever kind can be fed from the alleys conveniently and in the least possible time. If desired a tank can be built over the work-shop in the wing, and water run to the feeding room and each stall, which will give the stock water at all times. This is better than to allow milch cows in winter to go to the watering trough in the barn yard, waiting to have the ice broken before they can get any water, which then chills their system and seriously checks the flow of milk. The partition between the horse and cattle stable should be made to separate them entirely. The harness room keeps the harness free from dust and ammonia, and if desired, a work-bench may be put in for repairing harness when there is not much else to do. There can be no excuse for allowing machines and tools to lie out in the fields or about the out-buildings, as there is ample room on the two floors of the annex. On the lower floor one can drive in

FIG. 1.—PERSPECTIVE VIEW OF OCTAGON BARN.

with large machines and wagons, unhitch, and allow the horses to go directly into the stable. In the drive-way to the barn floor an opening should be made, through which to unload roots right into the cellar. The latter being near the feeding-room they can be fed with no waste of time. Between the tool-room on the second floor and the work-shop is a wide sliding-door, which allows any machine or tools to be repaired or painted with little trouble. A stove can be put in during the winter. The granary being over the feeding-room

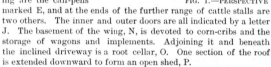

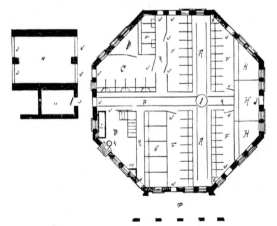

FIG. 2.—PLAN OF BASEMENT.

various kinds of grain and feed, and two shutes, B B, lead to the steam box in the basement. A hatchway, C, opens directly over the tramway and through it any feed may be dropped directly into the feed car below. The shutes marked E lead to the hay-racks of the horse stalls in the basement. A large ventilator, D, extends to the cupola on the apex of the roof and also serves as a shute for hay and straw. The doors on the floor are all marked F. In the wing, the large room, H, is for tools and small machinery, and G is a general repair shop with a chimney, L. A narrow corn-crib, K,

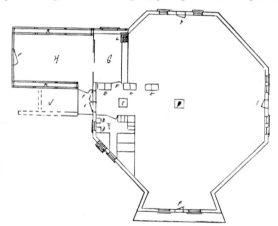

FIG. 3.—PLAN OF FIRST FLOOR.

and connected with it by shutes, all feed can be dropped direct into mixing-box. In threshing, the grain can be easily put in by one man."

The Tight Rein.—Most farmers who give no particular attention to horses usually drive with a loose rein. This is well enough with the "old family horse" in whom you have perfect confidence. It is never safe, however, with a young or spirited horse. Never drive such an animal with so loose a rein that you cannot instantly command the situation, whatever happens.

Europe are said to have traveled so fast because they lived on mare's milk and blood tapped from the necks of their horses. When Alexander the Great, who was no slouch of a horseman himself, conquered Persia he found horse breeding farms of as many as 20,000 mares maintained in the hills to supply the armies. In fact, horses reigned supreme in war and conquest until Alexander got to India where he ran into the war elephant. Later Hannibal of Carthage in North Africa sought to overwhelm the Romans by mustering an army on elephants, but the horse prevailed after the elephant returned to his native haunts.

We do not know just how or when the blood of the war horses worked its way down to the modern breeds, but the relationship is undoubtedly close. The Arabian is a distinct blood line highly prized today. Its hardiness, great stamina, speed, intelligence, and superb handling qualities probably can be traced to its development in the turbulent Middle East where a man's life often depended on his horse.

The big horses that were largely bred in Europe to carry the knights in armor are said to have been fighters as well as carriers of prodigious loads. The weight of the rider and armor (including the horse's armor) is believed to have been well over 500 pounds. Carrying this heavy load, the animals were expected to maneuver swiftly,

HERE IS THE
HARNESS
THAT WE ARE SELLING SO MANY OF AT
$5.50.

Send P. O. Order or Registered Letter and get a Set with Over Check or Side Rein.
WE DEFY COMPETITION.

LOOK HERE!
What the Public has to say about the **BARKLEY GOODS.**

NEW IBERIA, LA., June 7, '90.
*Dear Sirs:—*I have received the buggy in good order, and am perfectly satisfied with it. *I think it the most nicely finished buggy I have seen in this section of the country. Every one admires it.* MY WIFE IS HIGHLY PLEASED WITH IT.
I am very respectfully,
J. B. WINTERS.

GRANGEVILLE, IDAHO, June 16, 1890.
*Dear Sirs:—*The set of $5.50 harness that I ordered for Chas. Bentz, of this place, came O. K., and *every one here* WERE MUCH SURPRISED, AS THEY ARE AS GOOD AS HARNESS SOLD HERE FOR $20. Enclosed please find $10 for which please send your No. 6, $10 harness (nickel trimmed), by express to Frank Vansice.
Yours truly,
E. BECK, Postmaster.

MANCHESTER DEPOT, VT., June 30, 1890.
*Gents:—*The top buggy and road cart are at hand all O. K. They are VERY SATISFACTORY, and in every way appear fully equal to your representations. I think you may, in due time expect other customers from this locality, *as the goods are liked* BY ALL *who have seen them.* Yours very truly,
R. H. BRADLEY.

WE ALSO MANUFACTURE
GOAT HARNESS
from $1.50 to $16
and **GOAT or DOG CARTS**
at $4 and $7.
Write for
GOAT CATALOGUE.

For 20 consecutive years we have made and sold HARNESS to Dealers, BUT NOW we are selling direct to the consumers, saving you the traveling man's expenses and dealers' profits.

ROAD CARTS $11.50, BUGGIES $55.00 Upwards
Write for Illustrated Catalogue and Prices.
FRANK B. BARKLEY MFG. CO. 282 & 284 Main St.. Cincinnati. O.

ABOVE

HOW MANY FFA boys of our day know how to harness a horse? Or name the different parts of a set of harness? Here is a simple advertisement to practice on. Today's horse fanciers will find their mouths watering as they note the prices for harness, road carts, and buggies back in 1885.

OPPOSITE PAGE

THE AUTHOR has words of praise for the bronco teams that did much good work on Midwest farms in the early days. This wash drawing is by Perry Carter, Chautauqua artist and newspaper cartoonist who wrote feature stories for the Worthington [Minn.] Globe when the author was editor back in the 1930s. Carter was *especially good at drawing horses. He did this sketch to illustrate the story of a courageous farmer with a team of broncos who volunteered to drive through the night to deliver a corpse to an undertaker 25 miles away. A violent thunderstorm came up and scared the daylights out of both man and team.*

NANCY HANKS—QUEEN OF TROTTERS.

L. F. CAGWIN.

A new and powerful impetus has been given to the interest in the affairs of the light harness horse; for the recent astonishing performances of the trotting mare, Nancy Hanks, have stimulated public sentiment to a degree that is not short of a widespread recognition of the appreciation of the wonders which may be done by the cultivation of the trotting gait.

As a road horse, the American trotter is destined to receive that attention which will insure the maintenance of his popularity and progress. To this end not only is it necessary to cultivate the harness horse, but also the establishment of suitable roads on which he may be used to advantage. This latter factor would lead to an enormous increase in the value of driving horses. Last year Sunol beat the previous record, that of Maud S, 2:08¾, by trotting a mile in 2:08¼. This, as accomplished on a kite-shaped course, was

Dictator is the sire of Jay Eye See, 2:10 trotting and 2:06¼ pacing. The second dam of Nancy Hanks was Sophy, also second dam of Adrian Wilkes, sire of Roy Wilkes, the great pacer, by Edwin Forrest. Nancy Hanks is, therefore, inbred; both her sire and her dam's sire being by Hambletonian. In appearance Nancy Hanks seems to favor the American Star conformation, or at least that branch of the Star element to which Dictator, Dexter, Jay Eye See and their near kindred belong. Last year Nancy was placed in the hands of the great reinsman, Budd Doble, who gave her a record of 2:09, shortly after which she became the property of J. Malcolm Forbes, Boston, by whom she is now owned. This year she started several times unsuccessfully, and it was not until she reached Chicago that Mr. Doble considered her in fit condition for a supreme effort, and, when she was fit, he did not disguise the fact, but boldly announced that the mare was very likely to beat all records. Accordingly, at Washington Park, Chicago, on Wednesday afternoon, August 17th, when Mr. Doble appeared, seated behind the

THE REIGNING QUEEN—NANCY HANKS.

not justly considered equal to the record of Maud S, which yet remains the fastest mile ever made by a trotting horse on an elliptical course and attached to a sulky with the ordinary wheels.

The mile of Nancy Hanks, 2:04, which was made at Terre Haute, Ind., September 28th on a regulation track eclipses all previous harness records. She was so evenly rated that she made the quarters in .31, 1:02¼, 1:32¼, 2:04. The pneumatic tires are of some help but this performances shows her to be the fastest harness horse that has yet gone from wire to wire.

Nancy Hanks is a dark bay, or brown, mare with near hind pastern white, and a touch of white on off fore heel. She is fifteen hands and three-quarters of an inch in height, and weighs, in trotting condition, 860 pounds. She was bred by Mr. Hart Boswell, Lexington, Ky. She was by Happy Medium, son of Rysdyk's Hambletonian and the great trotting mare Princess, the rival of Flora Temple. Her dam was Nancy Lee, by Dictator, brother to Dexter, 2:17¼, by Hambletonian out of Clara, by American Star.

charming little mare, it was with a smile of confidence, which was in no wise diminished when, immediately after the performance of the mile in 2:07¼, he announced that in twenty minutes she could easily be ready and capable, if necessary, of repeating the mile in the same time, or better. Two weeks later, August 31st, at Independence, Iowa, the wonderful little Nancy eclipsed herself and lowered her former record just two seconds, making her record the remarkable one of 2:05¼. Nancy Hanks is the third trotter that has acquired the distinction as reigning monarch of the turf, at the hands of Budd Doble. The others are Dexter 2:17¼, and Goldsmith Maid 2:14.

Nancy Hanks is powerfully made about the stifles. She has strong, broad and cleancut legs. She is very deep through the shoulders and heart, and, in trotting form, well drawn up in the flank, and her stifles reach below her flanks very conspicuously, a form usually seen in extraordinary trotters, but more pronounced in Nancy Hanks than is usual.

and get in some licks with their own teeth and hoofs.

I wish we knew more about these great chargers, their breeding, care and the extent to which their blood has carried on into today's draft breeds.

THE INDESTRUCTIBLE BRONCO

While we note that the Arabians have come out of the rugged life and fierce conflicts of the Middle East, and the heavy breeds of Europe have descended from the chargers that paced the Crusades, I want to put in a word for another strain of the multinational horse that played a leading role in the making of America. Certainly to horsemen in general and most especially farmers, the word "bronco" calls to mind a distinct type of horse, even though it is not recognized as a breed. The bronco was wiry, medium-size to small, tough, and ornery. This horse of the American West was probably derived from the Indian pony, the wild mustang, various racing and thoroughbred types that found their way west as army mounts, and

OPPOSITE PAGE

HORSE RACING enthusiasts who were pulling for the filly Ruffian in the spring of 1975 in her tragic attempt to oust the stallions from their supremacy will be comforted to know about another great mare, Nancy Hanks, who really did the trick back in 1892.

AT RIGHT

DO YOU REMEMBER that verse, "For want of a nail a shoe was lost"? A horseshoe nail, which is like no other nail, was important enough to rate this poetic advertisement in the 1887 issue of the New York Sportsman.

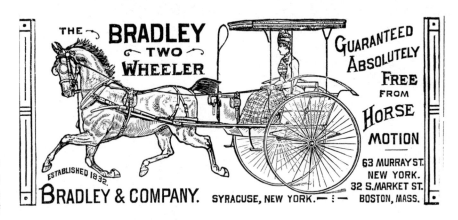

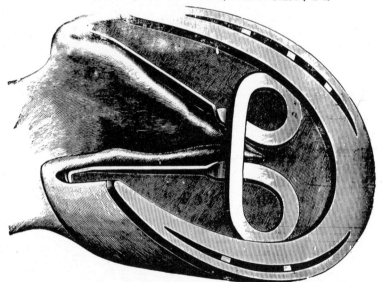
coach horses of English origin. While broncos did a lot of the work carrying the trappers and scouts, hauling covered wagons, and working the farm machinery of pioneer days, they were considered too light for really heavy work. Draft stallions—Percherons, Belgians or mixed heavy breeds—were in great demand in the late 1800s to turn loose with the bronco mares on the plains of the Dakotas and Montana. The result was a heavier farm horse (1200 to 1500 pounds) that took charge of the farm machines which turned the Midwest prairies into the "breadbasket of the world."

I had several uncles who took land in North Dakota in the late 1800s when it was believed that Dakota farming was just an extension of the way farming was done in Minnesota and Iowa. Scrounging for a way to get ahead, they made a business for a time of shipping carloads of these crossbred range horses to my father in Minnesota, who in turn broke them and sold them to neighboring farmers.

EARLY FARM HORSES

The Bronco-Percheron cross was a durable, general purpose animal which prevailed on farms when I was a boy. Farmers were later to get started with breeding heavy draft horses of purer blood—Percheron, Belgian, Clydesdale, and Shire were common breeds. They were in

AT LEFT

THE HORSESHOE did more than protect the horse's hoofs from bruising by hard and stony road surfaces. Horses, like children today, often needed corrective shoes—to straighten a "cocked ankle," to prevent interfering [hitting one leg with the other], or to alter a gait. The good blacksmith horseshoer knew how to change a shoe to deal with all these problems, especially if he did work at a race track.

demand for farm work and to pull coal wagons, beer wagons, and about everything else that needed moving as America began to flex her industrial muscle. We farm boys, who were all youthful horsemen, dreaded to see a good team go off to the coal wagons or the highway or railroad builders. It was like sending a man to Siberia. But the price was a cut above what could be realized from selling to farmers, so we shed a passing tear and went right on raising horses.

I cannot pass on to discuss other horse lore without saying a bit more about the indestructible bronco. Whether his prowess was due to the conglomeration of genes that went into this versatile little horse, or the tough environment in which he was raised, or the tough people with whom he associated, nobody will ever know. The bronco could be almost any color, black, bay, buckskin, roan, sorrel, spotted—you name it and it was there.

My Uncle Ernest, who ran a livery stable for a time in Cooperstown, North Dakota, drove mostly broncos. He loved to tell stories of the feats of the teams he owned. It was in the time of the diphtheria epidemic in North Dakota that killed literally thousands of farm children. Uncle had the job of transporting the doctor to remote ranches where the epidemic raged. He declared he could hitch his team of broncos to a bobsled in the dead of winter, drive 25 miles to a farmstead and return the same night. The exhausted doctor would be asleep in the wagon box on straw, covered with buffalo robes. The driver would be nodding on the seat, while the tired team jogged all night long at a speed which ate up the miles but somehow conserved strength. Uncle said that as they came over a rise and saw the lights of the town a couple of miles away, those tired broncos would break into a brisk trot.

AT RIGHT

A SAFETY PIN big enough to secure a horse blanket to a horse was a very useful item on the farm a century ago. It was too big to secure the baby's diapers, but it could be pressed into service once in a while to hold up a little boy's pants. Farmers who left horses unblanketed at a hitching post in the dead of winter were despised by more careful horsemen. On the way home the blankets doubled as comforters for the children snuggled down in the sleigh box. They smelled deliciously of horse!

For want of a blanket he won't bring fifty dollars.

Nothing keeps a horse in good condition like a blanket.

FREE—Get from your dealer free, the 5/A Book. It has handsome pictures and valuable information about horses.

Two or three dollars for a 5/A Horse Blanket will make your horse worth more and eat less to keep warm.

Ask for
{ **5/A Five Mile**
5/A Boss Stable
5/A Electric
5/A Extra Test }

30 other styles at prices to suit everybody. If you can't get them from your dealer, write us.

25

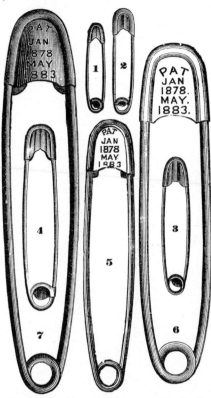

26

MULES and their smaller ancestors the donkeys [also called burros, asses, jackasses and many other names that are unprintable] are not given the emphasis they deserve in this book. Their contributions to the welfare of mankind are legion. Their popularity decreased as you worked your way north, so there were not many in the region where the author spent his boyhood. The upper photo shows a magnificent six-horse team of mules working on an Iowa farm. These are big mules, indicating that the jack who sired them, crossed with mares, must have been a big one.

The lower photo was taken in Latin America where mules and asses, along with oxen, were still extensively used for all kinds of hauling when the author visited these countries in 1956.

AT RIGHT

WHEN MULES INVADE horse country, as in the case with this team ready to take the Iowa farmer to town at a smart pace, they seem always to have been fitted with harness too big for them—like a growing boy stuck with his older brother's hand-me-downs. The mules never seemed to mind but went about doing their work, often more work for less pay than their horse relatives. Mules had a reputation for wilfulness and bad temper, not always deserved. The stories of their prowess and faithfulness are an important part of American literature.

A GOOD TIME TO LIVE!

I am glad that I have lived in the horse and buggy age as well as in the automobile age and had the chance to savor the experiences of both. One of the horse and buggy thrills to be cherished is the feel and acceleration of a spirited team as it took bit in teeth and headed for the home barn and a feed of hay and oats.

It is impossible to exaggerate the closeness of the relationship between man and horse in those days when they lived together and literally owed their lives to each other. Some say that the automobile today becomes an extension of man's personality, that a man at the wheel is a different person from the same man on foot. I don't hesitate to say that the affinity of a man and horse was much greater because the horse was a personality in his own right, much more so than a car can ever be. In the dovetailing—as often as not the clashing—of the personalities of man and horse, interesting things were bound to happen.

Man talk was largely horse talk in those days. Horses weren't exactly people, but they came very close to it.

My father, who couldn't get the habit of horse-trading out of his system even after we got to breeding purebred Percherons, knew every horse in the county. He knew them very much as he would know people, their ancestry, their good traits and bad, their work habits, their temperaments, even their metabolism. There were "easy keepers" and "hard keepers." The latter were inclined to be scrawny and untidy, no matter how well you fed them. That did not prevent them, however, from being good performers under saddle or in harness.

THEY KNEW THEIR HORSES

The vast knowledge that horse people managed to accumulate was something to be wondered at. Prairie Farmer columnist John Turnipseed has captured the flavor of the "horse culture" in a piece which I quote in part. The setting is that of two old timers on a bench on the courthouse lawn, noting and commenting between naps on the activity in the sleepy town:

Purty soon there is a gosh-awful racket an' down the street comes an old car spittin' fire an' screechin' its tires around the corner. Jeff opens one eye.

"That's the oldest Haynes boy. He got his pa's 1957 Ford an' he put a Mercury motor in it. He is runnin' around with one of the Schmidt girls an' her folks don't like it."

So we went back to snoozin'

THE ORANGE JUDD FARMER.

Devoted to Farming ; to Live Stock of all Kinds ; to Dairying ; Markets ; to Horticulture in all its Branches ; to Housekeeping ; to the Young.

Vol. IV, No. 15. { 308-316 Dearborn St. }   CHICAGO, ILLINOIS, OCTOBER 13, 1888.   { 20 Cts. to Jan. 1, 1889. }   One Dollar a Year.

CLYDESDALE STALLION "HOLYROOD" 3362, IMPORTED AND OWNED BY GALBRAITH BROS., JANESVILLE, WIS.

Holyrood (4446) 3362.

We present herewith an illustration, which in itself is a thing of beauty. The animal represented, "Holyrood" (4446) 3362, is owned by Galbraith Bros., the well-known importers and owners of English breeds of horses, Janesville, Wis. "Holyrood" was imported in 1887 by his present owners. He was bred by Alex. McCowan of Newton-airds, Dumfries, Scotland ; foaled May, 1884. In 1885 he won the second prize at the Royal Agricultural Society's Show ; third at the same Show in 1886, and third at the last American Horse Show in Chicago. His sire, "Auld Reekie" (1920) was the winner of the first prize at the Royal Society's Show in 1883, and other honors. His dam, " Kate of Banks," is one of the best mares of Scotland, a well-known prize winner at the Highland Society, and other important Shows. Of Messrs. Galbraith Bros., we need say but little, as they are so widely known. During the past year they have imported a large number of Clydesdale and English Shire horses, and have given the well-known Cleveland Bay a prominent place in their stud. They are also turning their attention to Suffolk-Punch Horses—a breed which will undoubtedly attain popularity in the near future. Messrs. Galbraith Bros. have had many years' experience in breeding and handling the best class of stock. They have a resident partner in Scotland, and above all, a life-long intimacy with all the well-known breeders in England and Scotland. For handling superior stock, and for fair and honorable treatment of customers, they are endorsed by every agricultural journal.

Sell Good Old Corn Now.

Yesterday (October 8th) No. 2 corn sold in Chicago for 44½@45 cents for cash and for October delivery ; at 45½ for November delivery ; at 41⅛ cents for December, and at 38¼ cents for January. A year ago to-day, when there were 550 to 600 Million Bushels less of crop than now, corn for October and November delivery sold at 42⅛ cents, and for December at 41⅜ cents. It would seem remarkable, at first view, that with so large a yield, the present price should rule 3 cents higher per bushel. But it is explainable, first, by the fact that little new corn will be ready for transportation and storage before November and December, and so there is a brisk demand for all old corn that can be used to fill contracts, or enter into immediate transportation for distant consumption. When the present crop comes freely to market, it would seem to be impossible for present rates to keep up, in the face of the immense supply, and the Board of Trade offers for future months, indicate a general expectation of lower prices, though the large wheat shortage will help keep prices higher

HOLYROOD, a Clydesdale stallion imported around 1888 by Galbraith Bros. of Janesville, Wisconsin, graced the front page of the Orange Judd Farmer in October of that year. And what a job the engraver did! The Galbraiths were importers of several species and breeds of livestock in the 1880s and 1890s. Scotsmen themselves, they had a weakness for the Scottish breeds. The Orange Judd Farmer has not been in existence for nearly fifty years.

BELOW

BEER WAGONS were a common sight in every city around the turn of the century, and somehow the beer companies started a fad of having the finest of draft teams. This photograph, kindly loaned by the Anheuser-Busch company of St. Louis, is of their famous eight-horse team of Clydesdales. Another beer company uses Shires in their exhibition teams. In fact, the beer companies have become one of the greatest preservers of good draft stock, using their teams of up to forty horses to charm the public at fairs and expositions.

an' I got to thinkin' how this courthouse small talk ain't what it used to be. Fifty years ago it would have sounded more like this if a team an' wagon was comin' down the road:

"Looks like old man Perkins is comin' to town for lumber an' a plug of tobacky," opines the one bench-setter. "That off hoss looks like the bay mare he traded off Charley Johnson for a gray mule an' the spavined roan with the blind eye. That there mare is part Hambletonian out of a stud over on the Kentucky side."

"Don't look like the Hambletonian to me," sez the next benchsetter. "She's got more of a star than a stripe. I reckon that ain't Perkins at all. Must be Johnny Burnside. That's the bay gelding he got off his father-in-law in a trade with the little chestnut he raced at the county fair two years ago. The chestnut got the heaves because Johnny was careless with dusty hay so now the old man is mad at Johnny an' sez he will cut his daughter out of his will."

"Nope," sez the first bench-setter. "Tain't the gelding. It's the mare. I can tell a mare by the ways she holds her head, an' besides she favors the left front foot with the wire-cut she got when she run away with the binder whilst the Perkins boy was settin' by an oat shock sparkin' the hired girl who had brung him lunch to the field. I knowed it was the mare!"

BOYS WITH BIG EARS

Listening to yarns about horses and their remarkable escapades was a standard entertainment when I was a boy. The pastime had a flavor that beats modern television all hollow. It was not limited to farmers. The horse was not just a country animal. He had thoroughly infiltrated the small towns and the big cities, although direct contact with horses in the cities was limited to the people who drove them. The livery stable was an important institution in every town, and a favorite hangout for the story tellers. The smell of horse, hay, manure, and harness joined to produce a piquancy which brought out the best in the story tellers. Many a small boy sat in an obscure corner drinking deeply of adventure associated with horses great and small, at the same time absorbing a vocabulary of four letter words which today's youngster can get only out of the best selling novels. Mothers did not approve of

THESE COLORED lithographs of the fabled Dan Patch, offered in 1912 by the owner, M. W. Savage, are treasured by collectors today. Until very recently this horse held the record for the fastest mile, 1:55, ever traveled by a horse in harness. M. W. Savage, who gave Sears Roebuck and Montgomery Ward a run for their money in the mail order business early this century, bought the horse in Indiana and gave him a home and indoor track at Savage, Minnesota. No horse would ever challenge Dan Patch, he was so fast! He always ran against the clock at fairs and race events. The M. W. Savage Company of Minneapolis could not hold its own against the Chicago mail order giants but continued as the International Stock Food Company for many years after the mail order business was closed.

MARGUERITE.

AT RIGHT

WHERE DID ALL the buggy horses come from that kept America on the move at a lively pace before the advent of the automobile? Many came from England, from where they were imported with considerable fanfare. Here are a couple of the best: top engraving, French Coach mare Marguerite imported by M. W. Dunham of Wayne, Illinois, in 1888; lower illustration, prize English Hackney mare Movement. These coach horses came in assorted sizes and were versatile. Some became army mounts and found their way out West where they were crossed with the inevitable bronco to produce some very durable horses that gave life to the "surrey with the fringe on top" and other light vehicles.

BUFFALO BILL'S HORSE CHARLIE.

WILLIAM S. CODY (BUFFALO BILL).

My gallant and faithful horse Charlie, which found a grave beneath the weltering waves of the Atlantic Ocean, was twenty years old at his death. He was a half-blood Kentucky horse, and was bought for me as a five-year old in Nebraska. From that time he was the constant and unfailing companion of my life on the Western plains and in the "Wild West" exhibition. He was an animal of almost human intelligence, extraordinary speed, endurance and fidelity. When he was quite young I rode him on a hunt for wild horses, which he ran down after a chase of fifteen miles. At another time on a wager of five hundred dollars that I could ride him over the prairie one hundred miles in ten hours, he went the distance in nine hours and forty-five minutes.

When I opened my "Wild West" show at Omaha, in May, 1883, Charlie was the star horse, and held that position at all the exhibitions in this country and in Europe, where I took the show in 1887.

billows you must rest. Would that I could take you back and lay you down beneath the verdant billows of that prairie you and I have loved so well and roamed so freely; but it cannot be. How oft at the most quiet hour have we been journeying over their trackless wastes! How oft at break of day, when the glorious sun rising on the horizon has found us far from human habitation, have you reminded me of your need and mine, and with your beautiful ears bent forward and your gentle neigh given voice as plainly as human tongue to urge me to prepare our morning meal! And then, obedient to my call, gladly you bore your burden on, little knowing, little reckoning what the day might bring, so that you and I but shared its sorrows and pleasures alike. Nay, but for your willing speed and tireless courage I would many years ago have lain as low as you are now, and my Indian foe have claimed you for his slave. Yet you have never failed me. Ah, Charlie, old fellow, I have had many friends, but few of whom I could say that. Rest, entombed in the deep bosom of the ocean! May your rest nevermore be disturbed. I'll never forget you. I loved you as you loved me, my

"CHARLIE," BUFFALO BILL'S FAMOUS HORSE.

In London the horse attracted a full share of attention, and many scions of royalty solicited the favor of riding him. Grand Duke Michael of Russia rode Charlie several times in chase of my herd of buffaloes and became quite attached to him. In May last, the English engagement having closed, we all embarked on the "Persian Monarch" at Hull for New York. On the morning of the 14th I made my usual visit to Charlie between decks. Shortly after the groom reported him sick, and I found him in a chill. He grew rapidly worse in spite of all our care, and at two o'clock on the morning of the 17th he died. His death cast an air of sadness over the whole ship, and a human being could not have had more sincere mourners than the faithful and sagacious old horse. He was brought on deck, wrapped in canvas and covered with the American flag. When the hour for the ocean burial arrived the members of my company and others assembled on deck. Standing alone with uncovered head beside the dead, was the one whose life the noble animal had shared so long. At length with choking utterance he spoke, and Charlie, for the first time, failed to hear the familiar voice he had always been so prompt to obey:

"Old fellow, your journeys are over. Here beneath the ocean

dear old Charlie. Men tell me you have no soul; but if there be a heaven, and scouts can enter there, I'll wait at the gate for you, old friend."

Whereupon Charlie was allowed to slide gently down a pair of skids into the water. The accompanying engraving is a lifelike portrait of Charlie when at the age of fifteen years.

Marketing Apples.—I have sixty acres of orchard coming into bearing. My plan is to assort my fruit into four grades—extra, first, second and culls. The extras must contain only the very choicest specimens and be put up in smaller packages than the first or No. 1, which will be put up in full standard sized packages and will contain nothing but fair sound fruit of uniform size, as near as may be, and always graded up to the same standard, marked accordingly, and shipped to reliable commission men. The third grade, or No. 2, will be good, sound fruit that is not up to the standard of No. 1, and will go to the evaporator, while the culls will go to the cider press and then into vinegar. By a strict, uniform system of grading and packing, I am not ashamed to have my name appear on the package.	GEORGE B. ARNOLD.

their sons hanging around livery stables.

But there were also more respectable places where a boy might learn to know a horse. All the bigger and more affluent houses in town had barns out back, with their attendant manure piles and stacks of hay. Many of those barns, or carriage houses as they were called, were truly magnificent, matching the homes in quality and architecture. They have now been remodeled into apartments which are much sought after in many cities. There were in town also the larger barns where the dray and coal teams, the ice and milk wagon horses, and other useful servants of man were cared for. There was a steady stream of country teams coming and going in town, with frequent stops at the blacksmith and harness maker.

THE PERCHERON HORSE -From a Painting by ROSA BONHEUR.

THE HEAVY DRAUGHT HORSE OF NORMANDY—From a Painting by ROSA BONHEUR.

OPPOSITE PAGE

HERE YOU CAN read it, just as it appeared in the American Agriculturist of 1888; Buffalo Bill's tribute to his great horse Charlie, buried at sea on the way home after one of the Wild West showman's tours of Europe. Horse people may wonder about Charlie's breeding. The artist made him look like a Thoroughbred, but sentiment calls for at least a little bronco blood out of the Old West. Charlie probably would have done well as a modern quarterhorse. One wonders also how many horses and other prized livestock were buried at sea in the course of transporting thousands of animals from Europe to America. Losses, which must have been quite common, no doubt broke the hearts of many an importer, as well as causing substantial financial loss. Is there a record somewhere of such loss, perhaps hidden in a letter or a story in an obscure weekly newspaper?

ABOVE

WHAT DID THE ARTIST-ENGRAVERS use for models when they turned out all these very good illustrations of great farm animals? The live animal was used in many cases; at other times a photograph. But quite a few of the animals were famous enough to have been painted in oils. The two above were sketched by an artist from paintings by the greatest of all livestock painters, the French artist Rosa Bonheur. Suitable credit is usually given. Oh for a chance to own the original painting, or even the original sketch by the copyist!

33

THIS IS THE HANDY "HANSOM CAB,"

Kentucky Break Wagon.

[PATENTED NOV. 29, 1881, AND AUG. 2z 188z.

JOHN V. UPINGTON & BRO., Lexington, Ky.,
Pa entees and Sole Manufacturers.
TO OWNERS, BREEDERS AND TRAINERS.

In offering this Wagon we are confident that it will meet with the approval of all in the future, as it has done in the past.

The most unruly colt or horse can be broke to harness quicker (without injury to himself, driver or wagon) in this vehicle than any made. Cannot be turned over; colt or horse cannot rear up and fall back; impossible to kick when properly harnessed.

Write for circular, price list and testimonials. Order from the following Agents when convenient:
J. H. FENTON, 211 & 213 WABASH AVENUE, CHICAGO, ILL.
J. T. McCLELLAN, BYRN MAWR, PA.

FULL PLATFORM SPRING WAGON, made by the Elkhart Carriage and Harness Manufacturing Co., Elkhart, Ind.—Price, $55.

THEY WERE CHARACTERS

On the farm, in town, or out on the trail, horses were regarded as characters. The literature of the time is full of accounts of the horse that ran its heart out to carry the doctor to the sick child, or saved its master's life in the heat of storm or battle. Horses could stick to a trail or a muddy road on a pitch black night when man was helpless. Horses brought drunken masters home from the saloons late at night and did everything but put them to bed.

My own memory has its store of character horses that I knew as a youth on the farm. We had a driving team named Cap and Colonel that had quite a hand in helping our parents bring up our family. Cap was a rangy bay gelding, out of one of the crossbred mares father had acquired and a Hambletonian stallion that made a brief appearance in

OPPOSITE PAGE

CLEVELAND BAYS, imported from the Cleveland area in England, were popular coach horses in America for a time, but they slowly lost their identity in the crossing and criss-crossing to produce horses for many purposes. Coach types were a kind of middle road between the heavy draft breeds and the lighter riding and racing stock. Coach horses had to have enough heft to deal with heavy wagons, yet they had to travel long distances in a trot without tiring. Here are shown some Cleveland Bays of the 1880s. The stallion at top was imported by George E. Brown of Aurora, Illinois. Below is a Cleveland mare, Ruby, with yearling and suckling colts, imported by Wm. Fields and Bros., Cedar Falls, Iowa.

AT LEFT is an assortment of vehicles that were whipped around in great style by driving horses such as the Cleveland Bays. The Hansom cabs were quite common in eastern cities, following a style set in Great Britain, but this kind of cab never made much headway in the Midwest.

35

ABOVE

*THIS FRENCH DRAFT STAL-
LION, Faison, owned by C. Fairfield
and Company, Waverly, Iowa, importers in the 1880s, was one of the
sires imported from France that carried the French Draft banner before
it gave way to the better known
Percheron. It is not always easy to
tell the draft breeds apart, even for
the expert. French Draft horses and
Percherons are said to have been developed to pull the heavy stage-
coaches that were used on a large
scale in Europe. These horses were
expected to carry their loads over
very bad roads, mostly in a brisk
trot. In America they became the
most popular draft breed among
farmers.*

OPPOSITE PAGE

*THIS MAGNIFICENT Percheron
stallion, Brilliant, whose portrait was
painted by Rosa Bonheur before he
was exported, became the property
in the late 1880s of M. W. Dunham,
Wayne, Illinois.*

36

the community. Cap was one of those too-smart horses who could open most any gate or barn door. He had a streak of mischief in him and was probably the chief instigator whenever he and Colonel decided to stage a runaway, which they did every year or so even after they had grown old enough to know better. We came to the conclusion that they would get bored with routine life and stage a runaway just for the hell of it. They managed to smash up a few buggies—and once the backhouse got in the way—but that is another story.

Among the remarkable things that Cap knew was how to tell time. When working in the field, he would never start a new round after 12 o'clock noon or 6 o'clock in the evening. For years we racked our brains to figure out how he did it. One autumn day I was plowing with five horses strung out, Cap and Colonel as the lead team. The air was cold and the wind was just right so I was able to hear faintly the factory whistle on the milk plant 13 miles away. It dawned on me that Cap had much better ears than I. He could hear the whistle any day of the year.

A SENSE OF SPORTSMANSHIP

The remarkable thing to me was not that the horse had better hearing than I, but that he would always finish a round after the whistle blew before he would start acting up to show his displeasure at working overtime. I should explain that a "round," as we called it on the farm, was a full turn down the field and back with whatever implement was in use, starting from the end closest to the farmstead. Even if the round was just beginning as the whistle blew, the horse somehow had the idea that finishing the round was the right thing to do.

Horsemen tell me that this "sporting sense" is not uncommon among intelligent horses. They also have a sense of humor, as I can attest. How else would you explain why a staid old mare, wise in the ways of the world and expending no energy without good purpose, would make like a green colt and shy at a piece of paper when there was a timid and unsure person on the lines? She couldn't resist the temptation of "scaring the pants off" the green driver. You might apply the same logic to the ranch horse, long broke and steady as a rock, that bucks a couple of times just to shake up the cowboy a bit before settling down to the day's grind.

I have used the term "man and horse" so often in this rambling discourse that I had better stop right here to appease the spirit of wom-

R. Bonheur

37

King of the Valley

en's lib that is abroad in the land. Of course there were and are excellent horsewomen, not only handlers of what we call pleasure horses, but of farm teams and driving horses. Nevertheless, the sagas of man and horse as they unfolded through American history were from a man's world. Often woman's involvement was more a matter of competition with the horse than empathy with the animal. Men loved their livestock to the point where wives often felt they were playing second fiddle.

While I am making apologies, I may as well explain why there is so little in this book on race horses and the whole range of pleasure horses. My experience has been only on the edges of this particular horse world. I must leave the telling of the race horse story to someone better equipped. Undoubtedly the history of racing stock, and special performance horses such as the Lippizaners, is replete with exciting accounts of great horses doing great things. However, I have included some engravings of famous racing horses, with special emphasis on Dan Patch, one of the greatest. It so happens that Dan Patch was an idol of all horsey youngsters when I was a boy. Our farm was no great distance from the enclosed track which M. W. Savage, owner of Dan Patch, built for this fabulous pacer near the small town which became known as Savage, Minnesota. One of our principal means of transportation was a small railroad named Dan Patch after the famous horse (now known as the Northfield Southern Railway).

HORSES WERE WORKERS

Practically all the horses I knew as a boy worked for their living. It is hard to do justice to the variety and importance of that work. Railroads on the land and work boats on river and canal took over much of the long distance hauling as the 1800s progressed. Steam power nibbled at farming for a time. The internal

ABOVE

THE PERCHERON STALLION in the upper engraving is Gladiateur, owned in 1888 by Leonard Johnson and Son of Northfield, Minnesota. The sire in the bottom engraving is King of the Valley, an outstanding *individual in a rather rare breed, the English Draft, no close relation to the French Draft. This breed did not have much of a following in the United States, and the few individuals that had an identity to begin with were probably absorbed into the Percheron lines.*

combustion tractor gained ground from 1900 on, but it was not until after World War I that mechanical power began seriously to displace the horse. Up to World War I horses and mules did most of the work, on and off the farm. They lived on grass and hay, oats and corn, requiring something like 60 million acres to feed them. When we shifted those acres to other farm production and began to depend primarily on petroleum for fuel, there was a profound change in farming, and for that matter in the national economy.

Forbodings of the time when horses would be displaced by other means of power surfaced early among horse enthusiasts. The following was printed in *The Cultivator* of February 1841 in the form of a letter to the editor:

"In this age of canals and rail-roads, the utility of horses is materially diminished; and should predictions be verified it will soon sink much lower, by the substitution of the steam engine for the horse in field labor. . . . To me the proposition is far from being a grateful one. . . . What pleasanter picture does the whole round of rural avocations present us, than that of the farmer driving his sleek bays to the field on a bright spring morning, fastening them before a No. 5 whose moldboard gleams like burnished silver, and then turning those long straight furrows, which one cannot help fancying yield better crops than crooked ones!"

BUILDING THE BREEDS

The building of the breeds and bloodlines of horses progressed steadily through the 1800s and into the 1900s. There were few sharp breed lines in the development of light horses for racing, riding, and coach work. The breeders thought it proper to get characteristics they wanted where they could find them.

ABOVE

THE BIG BARN shown above was made ready in 1889 at Chicago for a Percheron show that drew animals from several countries. Judges were appointed by Ministers and Secretaries of Agriculture in France, Canada, and the United States.

There were main lines, such as standard breds, thoroughbreds, Morgans, Arabians, and many more, working on down to today's Quarterhorses. Known bloodlines were always present and much talked about, but the combinations are blurred. Registration associations developed their own flexible rules to deal with a conglomerate situation. I will not try to sort them out here.

It should be pointed out, however, that today is an age of pleasure horses, and the future undoubtedly belongs to these lighter breeds. Their antecedants are just as noble and romantic as those of the heavy

IMPORTERS were the elite of the horse world during the Golden Age of livestock breeding, roughly the last quarter of the 1800s. Here are just a few among the many who did a land office business in the Midwest and who made a great contribution to livestock improvement.

40

BELGIANS gained rapidly in popularity after the turn of the century and established themselves along with Percherons as the most popular American breed. From their origins in the Flemish lowlands where they started as war horses and later became solid farm horses, they carried on on this side of the Atlantic. Even though they were a little slower than the Percherons, they were bigger and some believed more reliable.

breeds with which I am more familiar.

The draft breeds have a more orderly history. Their development impinged more directly on farming, plus other kinds of heavy hauling which persisted until the motor truck took over.

We thought of all horses, even show horses, as work stock. Horses were raised mostly on farms by farmers. There was a lot of work to be done on all farms. The royalty of the horse world worked alongside those without pride of pedigree. There were a few specialized horse establishments that bred and exhibited horses with little thought to their usefulness. Also there were the elite among the elite, the importers, who were spoken of with awe by ordinary horse people.

HORSES BECAME HEAVIER

Rank and file farmers were primarily interested in better work stock. It stood to reason that heavier horses would do a better job with the heavier machinery that was being invented. Then there was the off-farm market which paid well for quality draft stock.

Most of us got our first introduction to good purebred horses through the stallions that were "on stand" in the community. The owner was often a fair to good stockman who picked up extra income by keeping a stallion at stud. When this situation began to get out of hand, most state legislatures established pedigree and health requirements as prerequisites to obtaining a license to keep a stud.

BRILLIANT

41

OPPOSITE PAGE

THE SHIRE, close relative of the Clydesdale but somewhat larger and more inclined to take to farming rather than heavy coach work, enjoyed a brief popularity in the United States—and no doubt deserved every bit of it. The breed still has its ardent followers, including the beer company that chose the breed for its exhibition team. This stallion is Ben Blythe, owned by Galbraith Bros., around 1890. The breed originated in England, very close to Scotland, and is sometimes thought of as a Scottish breed.

Even so, there was not a great deal of choice for the average farmer. If the stallion man chose to keep a Percheron, you bred your mares to him. If the choice was a Belgian or a Clydesdale, that was what you used. The idea was to get better and heavier farm horses.

In those days the stallion parading through the community was a sure sign of spring. Mostly the stallion man would ride in a road cart drawn by a pony, leading the dignified sire on his rounds. When the stallion spied the mares working in the fields, he would issue his ringing challenge that could be heard for a mile. It was rumored among us boys that the studhorse got twenty-four raw eggs a day with his oats.

When the stallion arrived on the premises the mares were brought in from field or pasture to be "tried."

This occurred behind the barn—out of sight of the house—with a "snortin' pole" separating the mare from the stallion during the teasing period. Little boys were chased off, lest they be hurt or lest they learn too much too soon about the facts of life.

Later, when we acquired pure-bred Percherons and had our own stallion, offering service to our neighbors, I got used to the noisy procedures. I even had to handle the stallion when there were no older men around. It was dangerous work.

HEYDAY OF IMPORTERS

Importers of draft horses, who enjoyed their heyday in America from around 1879 to 1900, were pushing Clydesdales from Scotland; Shires, Suffolks, and English Draft breeds from England; Belgians, Percherons and French Draft breeds from France and Belgium. Along with the heavy draft types they also imported, on a smaller scale, coach and road horses. It is hard to distinguish where the draft horse left off and the lighter coach horse began. As was so often the case, horsemen usually placed the need for the right kind of a workhorse over the purity of bloodline. Coach work ranged from somewhat heavy loads, as in the case of the stage coaches, to light buggies, carts, and buckboards.

One of the coach types most often imported was the Cleveland Bay which came from England, but there were many other bloodlines, if not breeds, in this middleweight category.

Percheron: This breed which emerged as the most popular in America got its name from the old French district of *La Perche* to which it is native. It became rather widely distributed in France and even in other countries. Breed names such as Norman and French Draft kept turning up as closely related to the Percheron in American horse business. These names seem to have faded away except among historically minded purists who persist in identifying more precise bloodlines. While the Percheron had much of the same Flemish blood as the Belgian, it may also have had blood of the Arab and other coach types. It was bred largely for stage coach work. The Percheron is considered quicker, somewhat more spirited, and inclined to do better in a trot than the other heavy breeds. Also it has a finer head and smaller and tougher feet for road work. The color is most commonly shades of gray, black, occasionally bay or sorrel.

Belgian: Belgians ranked second on the popularity scale in our part of the country. As might be expected from the name, this horse came from Belgium and was a descendant of the old Flemish heavy horse which was held in high repute during the Middle Ages as a charger for use in war. The Belgian became a horse for heavy hauling and farm work. It is a big animal, sharing with the English Shire the reputation for great weight. Stallions go well over a ton. The head is coarse and larger than that of the Percheron, and the hair on the legs longer with more "feather." In temperament the Belgian is considered somewhat sluggish. Most common colors are bay, chestnut, and roan.

Shire: This English breed is said to be descended from the old English Great Horse or "Black Horse" which carried knights in armor. It was therefore counted on to be agile while carrying a very heavy weight. The Shire's lineage traces to pre-Roman horses with some infusion from the large Flemish. When knighthood lost its flower, the Shire became an agricultural horse and was dominant in all England except for the eastern and northern counties where Suffolks and Clydesdales were in greater use. In spite of its size, which is equal to

SHETLANDS ABROAD AND IN AMERICA.

Drawn and Engraved for the American Agriculturist.

or greater than the Belgian, the Shire trots readily. In both England and America it became popular, along with the Clydesdales, for pulling beer and coal wagons. Bay and brown shades replaced over the years the traditional black.

Clydesdale: This is a Scottish breed from Lanarkshire. It has in it much the same blood as the Shire, going back to the Flemish and the English Great horses. It is as tall as the Shire, but appears leaner and longer of leg, and is less massive in build. The head is more refined and the neck more arched. Clydes are also considered livelier and more mettlesome than the Shires, although this could be a matter of opinion. This breed has the same hairy legs but if anything the hair is shorter and more inclined to the back of the leg. Most people would have trouble distinguishing between these two breeds.

One prominent beer manufacturer in America uses Shires in its fabulous eight-horse team and another uses Clydesdales. Mostly the colors are bay and brown, but almost any variation may turn up. The Clydesdale people got their breed association going as early as 1877, and there is great loyalty to the breed. While there were many importers, the Scotsmen Alex Galbraith and Sons of Janesville, Wisconsin, are credited with pushing the Clydes in this country, getting their imported stock from both Canada and Scotland.

Suffolk: This breed had a brief flurry of popularity in America but did not catch on. Suffolk must be the only county in England which has breeds of horses, cattle, hogs, and sheep named after it. The Suffolk horse is a solid and sensible farm horse, noted for "hanging in there" when it comes to pulling heavy loads. Its heavy body is long and low but legs are somewhat slender. Its origin has not come in for much research. The color is usually chestnut.

HERE ARE TWO rather obscure breeds that the up-to-date horseman should know about—just in case he might encounter them in a crossword puzzle. Top sketch is of a Norfolk horse named Phenomenon, probably related to the Suffolk. Lower sketch is of the English Cart Horse, Enterprise of Cannock, who won some kind of a championship in England in 1883. Lots of breeds were used as cart horses, but there may have been some kind of multibreed competition to determine what kind of horse was best for cartage on London streets.

JUST IN CASE the reader is getting tired of engravings and artist's sketches of famous horses, here is a photo of a winner at a modern show. The horse is unknown, but some reader may be able to identify it. The livestock photographers, some of whom became quite famous and were in wide demand, took over where the engravers left off. They practiced many of the same tricks of making animals look even better than they really were. The author thinks the engravers did the best job.

A COMMON GOAL

Engravings of draft horses shown in this book tend to make the breeds look alike to those of us who are not up on the fine points. The practiced breeder with his likes and dislikes could see all kinds of differences. Still, there was a common goal, including both beauty and usefulness. The goal was well described by stockman and showman R. B. Ogilvie, one of the founders of the International Livestock Show in Chicago:

Utility in a draft horse means absolute soundness, intelligence, a willingness to work, wearing qualities, and the ability to move large loads at a long easy stride. Accessories to these desirable qualities in a drafter are oblique shoulders, short backs, deep ribs, long, level quarters with heavily muscled thighs extending well down to the hocks, shanks of ample size and quality, pasterns properly set, and strong shapely feet.

These brief sketches do not begin to tell the whole story of the development of the breeds on the continent and the transfer of the tradition to America where enthusiasm for good horses found fertile soil among farmers and fanciers. There was plenty of work for the draft horses in the opening of the West. But it wasn't all work. The livestock tradition took root quickly, the county and state fairs were organized even before the Civil War. The passion for practical improvement became a part of rural culture. All breeds of livestock and poultry, along with fruits, vegetables, and women's fancy work, were exhibited with pride. The fairs were important

Sire for Road Horses.

The best horses in the world for light service are bred in the United States; indeed, it may be said that this country yields a better class of horses for all work than does any other on the globe; but, beyond all question, our road horses are signally superior to those of any other land.

There is money, good, substantial money, nor that in small quantity, in breeding horses well, with a fixed end in view. Every farmer has, or may have, brings, even when young, and the condition of the markets in cities and large towns, where good, well-looking horses, having good action, style, and substance, always are in demand, and they will conclude that it pays to breed well. With the foregoing ideas in mind, we present herewith the portrait of a horse possessing the most desirable qualities as a sire, Lawgiver, a Kentucky-bred stallion, owned by Mr. Thomas Broadfield, of Rome, Oneida Co., N. Y. He is a magnificent bay horse, with black points, except one white heel, behind, the turf, which is not near-of-kin to Lawgiver. His warm blood, derived from Messenger, Diomed, Sir Archy, American Eclipse, Saltram (son of English Eclipse), imported Monarch, Sir Harry, Bedford, Pantaloon, Expedition, and their fellows among his ancestry, gives him the faculty of imparting unusual vigor and endurance to his offspring, which will inherit, besides, that speed predecessor, Rysdyk's Hambletonian, Mambrino which is the golden gift of Lawgiver's immediate Chief, Alexander's Abdallah, and Curtis' Hambletonian.

THE TROTTING-BRED STALLION LAWGIVER

Drawn (by Forbes) and Engraved for the American Agriculturist.

a good mare; it may be not one of great intrinsic value, but a mare of good points, well-spirited, good on the road, well-modeled as a dam, with room to carry and ability to nourish her foal. This mare is capable of doing good service in breeding, equally with that she renders in other labors for her owner.

The farmer's aim should be the production of the best foal possible, because it costs no more to raise a good than a poor colt, and there is no economy in using a cheap or convenient stallion, when a first-class one, of fit blood-lines and high quality, can be had for a little extra trouble and a little more money. Let our readers estimate the value of care in breeding by considering the prices which good stock and a delicate white speck on the forehead. He was bred by Mr. T. J. McGibben, at Edgewater Stud, Ky., foaled in 1880; stands 16.2 hands, and weighs 1,201 pounds; he is strongly gaited, with fine knee-action; is lofty, finely crested, free, easy and powerful of movement, and well tempered; feet and legs perfect; without fault or blemish, and descended from the most vigorous strains of blood possessed by American horses—a fact which makes him prepotent in charging his get with his ancestral and personal characteristics.

There is no horse of great exploits on the trotting tracks of the country, from Dexter, the king of twenty years ago, to Maud S., the ruling queen of this is the kind of horse for farmers to select as a sire, if they are to look for money-producing colts and fillies. His character is fixed by his long pedigree, and he cannot help affecting his offspring, for such is the law of descent. There is no use in trying to breed valuable stock, if the breeder selects his sire at hap-hazard; breeding is a science, and there can be no sure outcome from either its study or application, unless the progenitor of the stock shall have been thoroughly bred. It is well to use a well-bred mare, whenever this is possible; but a strongly bred sire will unfailingly improve upon the most ordinarily bred dam, will more than make good her deficiencies, and compensate for the investment.

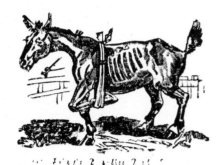

ABOVE

THE NAME ARAB on a horse is like sterling on silver. The top engraving shows rather well the characteristics of the Arabians which originated on the deserts of the Middle East long before people started to talk about oil. In this country there is some doubt whether "pure Arabian blood" can be found, but that does not dim the fame of these great horses. Most coach and driving stock, and even a heavy breed like the Percheron, is said to have some Arab blood in its background. The lower sketch here is of an English hunter, not a pure breed but incorporating the best of several strains of famous horses.

50

WALLACE'S MONTHLY

social events. Everybody got into the act, either as exhibitor or as observer.

At the great fairs importers did a thriving business. American breeders began to develop their own bloodlines. Horse talk flourished, and tempers flared as the virtues of the various breeds and strains were argued. A very good time was had by all!

The aura that surrounded imported stock was still evident when I was a boy. But it was about due for decline. Since the livestock of Britain and France was raised by peasant farmers as well as specialized breeders, you may wonder how the importers managed to collect their animals.

In Europe it was a common practice to form associations of smaller farmers who bought a superior sire and bred their farm mares which, undoubtedly, had to do a lot of farm work besides raising colts. There were collectors at the European end who kept track of colts coming along and took American importers

AT LEFT

HAMBLETONIAN is the name of a famous competition of harness racers held each year at DuQuion, Illinois. It is also the name of a famous horse foaled more than 100 years ago who was the beginning of a strain of trotters that has been considered among the best in American horse history. The engraving is that of Rysdyk's Hambletonian from the 1880s. When the author was eight years old, a bay stallion of Hambletonian breeding had a brief stand in the community, and father bred one of the medium-sized black mares to this stud two years in succession. The result was two bay driving horses, Cap and Kate. Cap stayed with the family well into the automobile age.

The lower engraving is that of an Exmoor, an English strain that few have ever heard about.

out to the small farms to bargain for stock. I have heard descriptions of how the buyer would sit down with the entire farm family in England or in France to make his purchase. Selling a young horse for export was an event which did not occur every year. Lasting friendships were nurtured across the sea, and the language of the stockman developed an international flavor.

The animals had to be assembled at a seaport and shipped. We know that many an immigrant to this country earned his passage on a "cattleboat." These livestock frieghters handled precious cargo from time to time, so you can be sure the livestock often got better accommodations than the people. However, it must have been a rather tricky trade, with some losses inevitable. Many an importer slept with his animals to insure their safe arrival.

There are still a few draft horses of superior breeding left, both in Europe and in America. The work-house suffered the tragic blow of having his work taken away from him by motor car, truck, and tractor. Preservation of the draft horse is now in the hands of a few horse lovers. Draft stock virtually disappeared from the fairs some years ago. There has been a revival lately but it does not have sound underpinnings—unless we really run out of gasoline. For a time the horse herds eked out a living by the sale of PMU (pregnant mare urine) to drug companies for the manufacture of estrogen for use in human medication. But even this product was synthesized.

DEPEND ON HOBBY PEOPLE

Draft breeds now depend almost entirely on the loving subsidy of the hobby people, plus the beer companies that still earn the ahs and ohs of the public with their beautiful hitches of matched animals, up to forty in number in a single hitch. That is the more reason why the reader should savor the beauty of the engravings shown here and enjoy the echoes of the Golden Age which they portray.

Pleasure horses are another matter. They are on the increase today as we try to recapture the spirit of the Old West and relive in a small way the "simple" life of our grandparents. The migration of commuters to the outer suburbs and onto acreages in the country made additional room for the pleasure horse. There is horse talk again in rodeos and on trail rides. This time the horse rubs up against a different kind of human master and mistress with different motives. The interest in breeds and bloodlines as applied to lighter horses is growing. New "breeds"—if they can be called that—are being created, for pleasure and profit. Too often the breeding is with an eye to color and markings, rather than spirit and durability, but that may change as the new breed of "horseperson" gets more deeply involved with the new types of horses.

There is an unhappy note, however, in this picture. Veterinarians have told me that the pleasure horses are not, by and large, being well taken care of. Often the love affair with a horse is a fleeting thing for young people in an age of hot-rods and fast travel. Many good horses are being neglected.

Horses may have worked harder and suffered more in the days when they were essential to the building of a new country. But they belonged. They had work to do. They were full partners with man and deserve a more prominent place in the history books.

AT LEFT

STANDARD OIL COMPANY made harness oil and thereby served the horse before automobile fuel became its chief business. Goodyear made rubber horseshoes and hard rubber tires for buggies and carriages before the firm really got going on the auto tire business. The first tires for cars were made of hard rubber, adapting much of the technology that had been worked out for horse drawn vehicles.

THE SMALL AD at the left is a hand-cranked machine for clipping horses, with a shearing head not too different from the clipper used by barbers today. In cold climates horses grew shaggy coats of hair to keep warm in winter. Often the coat had not been shed by natural means before the horses had to undertake spring work on the farm in comparatively warm weather. This called for a clipping job along about the first of May. Some farmers owned their own clippers, but often an itinerant would call at the farm with his equipment. He would clip a whole horse for fifty cents, with the boys of the family clipped free. Same machine same method! Young boys would appear in school with their heads bald as billiard balls even before the last snowstorm had passed. Mothers objected but fathers were indulgent, and there was nothing that set a boy up with his peers like a complete trim with a horse clipper.

CHAMPION DELIVERY WAGON.

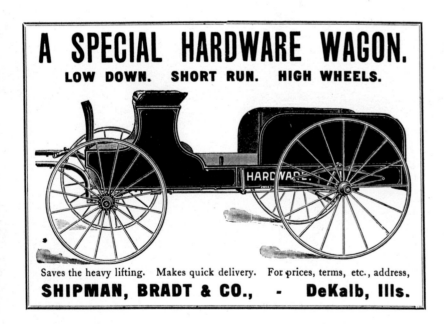
THE HORSE AND BUGGY age spawned a whole family of curious vehicles that would be the life of the parade today if you can find and restore them. Delivery wagons and cabs were especially interesting. Delivery to the home of practically everything that could be bought in a store was standard procedure. Suitable wagons were designed by ingenious entrepreneurs. Most older city people today remember only the milk wagon and the ice wagon, perhaps also the coal wagon, all of which made frequent calls at the city home.

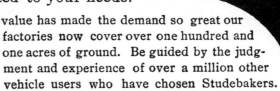

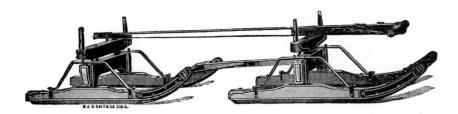
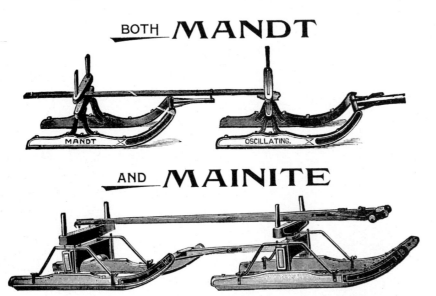
AT LEFT

HOW DID YOU GET AROUND in winter during the horse and buggy age? The answer is that some of the best traveling was done in winter by sleigh. Sleds were a good deal simpler than buggies and wagons. The cutter, pictured above, could be pulled by one or two horses—all you had to do was change from a pair of shafts to a single pole hitch. Cutters came in all sizes, from Spartan simplicity to an upholstered job with doors, and even a top. They were sporty looking but tricky to drive, overturning at the drop of a hat or the shy of a horse.

The good old bobsled was nearly always the rig of choice in winter. It was low, it was simple, and you could install on it any box or rack you used on a wagon. Horses could handle enormous loads if the snow was just right for sleighing. Usually the family traveled in a wagon box installed on a bobsled drawn by a lively team. The bottom of the box was covered with straw. The children would snuggle down crosswise with their backs to the wind, whereupon they would be covered with horse blankets, warm from the horses' backs, buffalo robes, or quilts borrowed from the house. It was a beautiful way to travel, especially with the sound effects of the sleigh-bells that were somehow scrounged up by the poorest of folks. Father would sit on a bench near the front of the box. Sometimes mother would sit with him, but if the wind was really sharp, she would join the kids down in the straw.

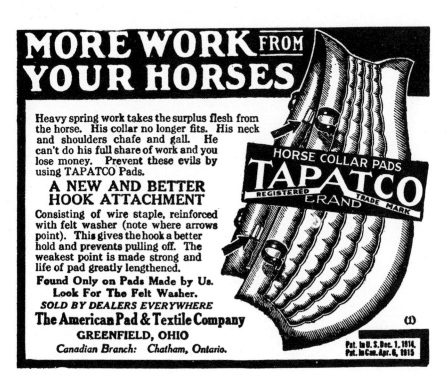
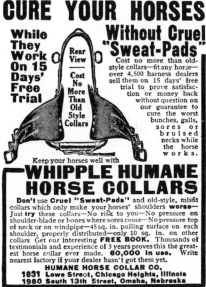
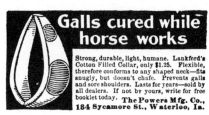

AMERICAN AGRICULTURIST

FOR THE

✦ FARM · GARDEN · & · HOUSEHOLD ✦

"AGRICULTURE IS THE MOST HEALTHFUL, MOST USEFUL, AND MOST NOBLE EMPLOYMENT OF MAN."—WASHINGTON.

VOLUME LII.—No. 11.   NEW YORK, NOVEMBER, 1893.   NEW SERIES—No. 562.

AMER. AGRI.

A. Bennell

THE FAMOUS SHORTHORN PRIZE WINNER, ABBOTSBURN.

In my discourse on horses I said that the horse was in on practically every important event in recorded history but never got much credit. The poets and storytellers did fairly well by him, for they sensed the romance inherent in the man-horse relationship; but the historians allotted him only a foot-note or two. Until recently historians have been by and large a shallow bunch, concerning themselves much with dates, battles, alliances and the coming and going of governments. They don't always tell what really went on.

They have precious little to say about who fed the philosophers, sculptors, and warriors of Greece; the conquerors and bureaucrats of Rome; and the squabbling factions of Britain who trampled the farmers' fields as they played tug-of-war back and forth across British soil. Behind statesman, artist, or soldier there has always been a peasant farmer (slave or free) who extracted from the soil enough food to feed his family plus something extra, and who husbanded the beasts on which civilizations were built. Buried in unwritten history are the stoical labors of the farmers who upheld governments, armies, and civilizations, working under great handicap. Their crops and their stock often were stolen or taxed away, their daughters ravished, and their sons impressed into military service, but somehow they got their job done. The politicians and soldiers received their needed rations, and the cumbersome rise and fall of civilizations continued.

Farmers of today who think they have it bad might give a thought to their counterparts of centuries past.

EVEN LESS CREDIT

Consider the plodding bovines—cattle to you and me. The Bible and other early historical documents include sheep, goats, swine, and other property under the term "cattle," but we have pretty well limited it to members of the *bovinae* or ox family. Like the horse, the

59

bovine was on hand when history was being made, but got even less credit in the history books.

The cow (and the bull) supplied a great deal of what man needed in war, work, and play—enormous amounts of power, meat and tallow, milk with all its component products, hides, horns, hair, and even manure. Sport and drama may also be added, such as the bull dances of the Greeks and the bullfights of the Latins. All this and more was contributed on a ration of grass and forage, with very little need for the concentrate grains that can succor man as well as beast.

There are fossil remains of the ancestors of our cattle going back for three million years. A million years ago there existed several forms of the ox, along with his cousins the bison and the water buffalo. All are rather closely related. Before the Ice Ages *urus,* the great Auroch, had appeared in both Europe and Asia. The ice gave these beasts a bad time and moved many south out of Europe, along with the other larger species that ended up in Africa and Southern Asia, or lost out altogether.

After the Ice Age the Aurochs and some smaller hairy cattle that had remained on the fringes of the glaciers took over as the dominant Big Beasts of Europe.

AUROCHS WERE TOUGH

The Aurochs deserve special mention because they were huge, powerful, and wild. There is at least a literary record of their surviving in the wild state in Northern Europe as late as 1625. In crossing with smaller cattle, notably the Celtic shorthorn (England was "cattle country" even before the last of the glaciers had receded), they were considered the chief progenitors of today's European cattle.

Very recently a zookeeper in Germany made an attempt to "breed back" cattle to achieve the Auroch type, and with some success, but there was no way to know exactly what was the goal.

Some form of the term *urus* may have been applied to cattle even before the Romans. Writing in his Commentaries in about 65 B. C., Julius Caesar described wild cattle

THE CULTIVATOR
FOR THE FARM, GARDEN AND THE FIRESIDE.

COUNTRY GENTLEMAN

Combined Papers.
Fifty-Sixth Year.

ALBANY, N. Y., JUNE 17, 1886.

Country Gentleman.
VOL. LI---No. 1742.

PUBLISHED BY
LUTHER TUCKER & SON,
EDITORS AND PROPRIETORS,
LUTHER H. TUCKER } *No. 395 Broadway*
GILBERT M. TUCKER } ALBANY, N. Y

Associate Editor:
JOHN J. THOMAS, Union Springs, N. Y.

THE COUNTRY GENTLEMAN is issued Weekly, and is designed to include, not in name but in fact every department of Agriculture, Stock-Raising, Horticulture and Domestic Economy. Subscriptions may commence with any month.

TERMS.—To City Subscribers, whose papers are delivered by Carriers, $3 per annum. To Mail Subscribers, $2.50 a year, if paid in advance, or $3 if not paid in advance. Subscriptions less than one year, 25 cents per month.

AYRSHIRE COW LADY KATE.

THE HOLSTEIN-FRIESIAN REGISTER
Is Published the 1st and 15th of each Month. Subscription, $1.50 per Year.

It is the only paper in the United States devoted solely to the interests of Holstein-Friesian Cattle. It keeps its readers thoroughly informed as to Sales, Transfers, Records, Tests, and General Progress of the Breed. As it reaches Holstein-Friesian breeders in nearly every State in the Union, it is an excellent advertising medium. DUDLEY MILLER of Oswego, N. Y., (Associate Editor,) contributes to each number of the paper, represents it generally, and especially in the Eastern and Middle States. Your Subscription and Advertising Patronage is most respectfully solicited. Address

Holstein-Friesian Register, Terre Haute, Ind., or Dudley Miller, Oswego, N. Y.

from the Hercynian Forest in Germany that were undoubtedly a strain of the Aurochs:

"There is a third kind of these animals which are called *uri*. In size these are little inferior to elephants, although in appearance, color, and form they are bulls. Their strength and their speed are great. They spare neither men nor beasts when they see them."

In speaking of bulls Caesar was, of course, referring to cattle already famous in sports of the Greeks and Romans, and to domesticated oxen which had been serving man for many centuries.

Domestication of the ox occurred near the end of the Stone Age. From then on, man and ox were inseparable. A person's wealth was measured to a large extent by the cattle he owned. Even before the time of Christ the ox was well established as the most significant power for both farming and heavy hauling. So well was this relationship recognized by Bible times that in the Apocryphal Book of Ecclesiasticus we find this poetic tribute:

*How shall he become wise that
   holdeth the plough,
That glorieth in the shaft of the
   goad,
That driveth oxen, and is
   occupied in their labours,
And whose discourse is of the
   stock of bulls?*

*He will set his heart upon turning
   his furrows;
And his wakefulness is to give
   his heifers their fodder.*

*All these put their trust in
   their hands:
And each becometh wise in his
   own work.*

*Without them shall not a city be
   inhabited,
And men shall not sojourn or
   walk up and down therein.*

One way or another, the *bovinae* have been modified by time and

ABOVE

HUGE STEERS weighing 3,000 pounds or more were quite the thing for a couple of centuries, culminating around 1850. They were prized as show cattle and sought for in meat markets. The biggest on record is said to have weighed 4,365 pounds. The steer shown above was one of twins, three-quarters Shorthorn, owned by the Hon. A. Ayrault of Geneseo, Illinois, in 1848 and shipped to New York City for slaughter. The pair weighed in at 5,522 and dressed out at 4,376. Reason for the popularity of the large animals: Tallow was worth more than meat because of its multiple uses in the pioneer economy. The steers described above were Durhams, a Shorthorn predecessor, slaughtered when five years old. Shorthorns or Shorthorn-Holstein crosses grew really big.

THESE WORKING OXEN were encountered by the author in 1956 in the sugar cane fields of Guatemala, Central America. The "company oxen," as they were called, were said to have been direct descendants of Spanish oxen. They tended toward a blue roan color. Most of the oxen in Central America were plain steers showing Holstein, Guernsey, or Shorthorn blood.

WHILE MANY BIG STEERS were grown only for meat and tallow, there were also those that got that way from years of service as draft oxen. Oxen were widely used for draft in America well through the 1800s. They are still in use in many parts of the world. The above illustrated pair of steers was owned by C. W. Barnum of Connecticut in the 1800s. They were three-quarters Holstein and one-quarter Shorthorn. Together they weighed 5,500 pounds.

BEFORE.　An Object Lesson in Dehorning.　AFTER.

THIS CARTOON published in an 1888 issue of the Orange Judd Farmer indicates there may have been an argument over horned vs. hornless cattle. Not all would agree with the implication.

BELOW

THE 1870 YEARBOOK of agriculture published this drawing of Texas cattle. The Longhorns were in that year finally finding their way to the slaughtering centers of the Midwest and the East in large numbers, thanks to the penetration of the West by the railroads. Texas Longhorns is a rather loose term applied to a wide variety of cattle raised in the Southwest, all the way from Texas to California. Many of these cattle were also trailed north to stock the ranges of Wyoming, Montana, and the Dakotas, where their wild nature was considerably subdued by crossbreeding with Herefords, Shorthorns, and other more sedate eastern cattle.

scattered over much of the world. Members of the family, with the exception of the water buffalo, will interbreed and produce fertile female offspring. It should be noted here that only one of the family seems to have gotten across the land bridge between Asia and North America; that was the American bison, which has been misnamed the buffalo. Our bison crosses with European cattle, although there are some difficulties to overcome in bringing this about.

VARIETY OF GENETIC MATERIAL

The versatility of members of the *bovinae* has proved in recent years to be a boon in supplying genetic material for experimental combinations of cross breeding. To begin with, breeders went to the Brahmans from tropical countries to get immunity from heat, ticks, and warm climate diseases. But this may be only a beginning.

One of the fears of the geneticists today is that widespread use of artificial breeding may cause our modern cattle to become too closely related and susceptible to being wiped out by new or newly virulent diseases. Today's technology in AI makes it possible for a superior sire to have 10,000 to 100,000 offspring. Furthermore, he can continue to inseminate females many years after he is dead. There is comfort in knowing that the ox family can offer a great variety of protoplasm from all over the world to be used in case of need.

WORKED FOR CONQUERORS

I cannot take time here to trace the whole development of the ox as a partner of man. Progress was undoubtedly pretty much uninterrupted as civilizations rose and fell. The ox may have had his "druthers" about whom he liked to work for, but he did not exercise his choice. There was work to do whether a

THE BEST KNOWN bull in America during the Golden Age of livestock breeding was no doubt Bull Durham, known intimately to farmers and cowboys as the makings for roll-your-own cigarettes. This handsome bull dominated the countryside through newspaper advertisements and billboards. The story is told that a Wisconsin dairy farmer sued the tobacco company because his cows spent too much time admiring the bull on the billboard and did not tend to the business of grazing and making milk.

civilization was waxing or waning. When the conqueror took over from the conquered, oxen and knowledge of their management were among the prized booties of war.

Judging from drawings and artifacts in the Egyptian tombs, cattle management was well advanced as early as 3000 B.C. There were cattle in England when the Romans arrived. Since these conquerors from the south were great civilizers and organizers, they set about improving those cattle. In fact, the Roman conquests and later, the Crusades, had a lot to do with introducing new blood from South to North, and vice versa, for improvement of both cattle and horses.

CLASSICS TOOK A HAND

I have complained about the historians being so preoccupied with dates and battles and royal families that they omitted to chart the bloodlines of livestock. Fortunately the poets and agricultural writers of both Greek and Roman periods did much better. The Greek Hesiod offered good advice on agriculture in his *Works and Days.* Virgil followed him later with his four books of the *Georgics,* offering tips on cattle management and other branches of agriculture. He gave us a pretty good idea of the state of animal husbandry in his generation.

This bit from Virgil's *Georgics* would be hard to beat as good advice in ranch management even today:

> *Next, when calving is o'er, man's whole thought goes to the offspring;*
> *And they stamp them anon with brands distinguishing each one*
> *As preference dictates; these for maintaining a true breed,*
> *These for sacred office, and these for laboring oxen*
> *Upturning the rugged loam-clods and straining across them,*
> *While the others at grass go forth as an army to pasture.*
> *[C.W. Brodribb translation]*

Other ancient writers whom you might wish to look into to learn about agriculture are Cato, Colum-

66

THE DOMESTICATION OF THE BUFFALO.

RICHARD WAUGH, MANITOBA.

The American buffalo or bison, *Bos Americanus*, worthily regarded as the "boss" quadruped of the Western Continent, which less than a quarter of a century ago could be found roaming over the great central plateau in countless thousands, is to-day on the verge of extinction. The few that are supposed to be preserved in the National Park at the sources of the Yellowstone are being stealthily shot down by poachers on that preserve, and the very sight anywhere else of a wild buffalo is promptly recorded in the newspapers. Now that the wild buffalo is gone or nearly so, a keen interest is felt everywhere in its possible domestication and reproduction, either pure or crossed with the common cow. C. J. Jones, known

is a circular limestone elevation about one hundred feet above the surrounding grassy prairie, on which stands the penitentiary, and where in the dead of winter the herd is collected and supplied with prairie hay. At any season they are quieter than the average of range cattle, but a haystack is a great help to their civilization, and

FULL BLOOD BUFFALO BULL.

CROSS-BRED BUFFALO STEER.

as "Buffalo" Jones, of Kansas, has lately with characteristic American enterprise gone eagerly into the collection and crossing of buffaloes, and has met with gratifying success. He has a herd of captured buffaloes and is breeding them to one hundred cows.

But the oldest and most successful demonstration of the possibility of domesticating the buffalo has been furnished by S. L. Bedson, the warden of Stony Mountain penitentiary, twelve miles northwest of Winnipeg, the capital of Manitoba. Stony Mountain

they spend a good deal of time in its vicinity, leisurely regarding their human visitors much the same as other winter-fed stock would. They seem to appreciate the blessings of farm life, when they come in the form of food and shelter. But they are what a Scotchman would call "kittle cattle" and a single stroke of their horn has been known to rip up a horse ridden incautiously too near them. An old buffalo cow has an evil eye at any time, and in the breeding season, when they retire from the open plain around the penitentiary to the poplar bush at the northeast, they have a very uncanny look which does not bear false witness against them. Precedence among the males is gained by pitched battle, but outside the breeding season they are usually on friendly terms, and three or four old toughs, sometimes a solitary one, will stroll away for weeks, paying little heed to the fences of the settlers. The bill for these chance sprees is promptly settled by the genial warden. On such a trip they are about as amenable to control as the leviathan or unicorn described in the book of Job, and a man sent in pursuit may have to use pistol shots to bring them to reason. One old bull that had the sinew of his hind leg cut by such a shot three years since may usually be seen at any season not far from home and is a grand specimen.

The Bedson herd was started in 1879, when Mr. Bedson bought five calves, a bull and four heifers, for $1000.

KETTON. 1st, (709).

CALVED IN 1805.—BRED BY CHARLES COLLING.

13.

DUCHESS BY DAISY BULL. (186).

CALVED IN 1800. BRED BY CHARLES COLLING. 10 YEARS OLD.

ella, and Pliny the Elder.

The poets of old sensed a certain amount of romance in the relation of the ploughman to his team. It may be hard for today's reader to understand this. What can be romantic about a couple of rangy steers working their heads off without raising a little Cain from time to time as horses and mules were wont to do?

OPPOSITE PAGE

THESE TWO stone lithographs by J. R. Page of Colling Shorthorns are taken from the book, History of Short-Horn Cattle, by Lewis F. Allen, published in 1874. The Colling family is credited by many with creation of the Shorthorn breed, beginning way back in the late 1700s. Symbolically this bull and cow might be called the father and mother of the breed.

SHORTHORNS have always done well in crosses, which explains why the King Ranch used this breed in crossing with the humped Brahman to get the modern breed, Santa Gertrudis, in an effort to get a beef animal able to take hot climates. The rare crossbred shown just above is between a Shorthorn and the Wild Chillingham cattle of England, thought to be the last wild cattle in Europe.

E.H. DEWEY DEL.

"Brant Chief."
CHAMPION SHORT-HORN. 1888.

69

EMPATHY BETWEEN DRIVER AND TEAM

Make no mistake about it, there was real empathy between a well-trained ox team and the driver. The team could be controlled by a gesture, the wave of a stick, a word or even a grunt. They could be taught to ease into a load and maneuver it in a way that few horses ever learned. They were less likely to get excited or unruly as horses often did when hitched to a load beyond their strength.

Competitive pulls have no doubt been held all over the world, but they became a major sport in America during the 1800s. Even today ox teams are being driven in parades and placed in competition by hobbyists who are eager to preserve an old art. The usual method of competition was to load a stone boat (a sort of skid) with rocks or cement blocks, which had to be moved a certain number of feet by a team to qualify. As the competition got keener, more and more stones were piled on the stone boat. The weight record for a six-foot pull was held in the early 1900s by a team of giant Holsteins which moved 11,200 pounds of granite blocks. It should be remembered that really big steers five or six years old might weigh over 3,000 pounds each.

MANY A BITTER ARGUMENT

There were long and bitter arguments over the relative value of horses, mules, and oxen for farming and heavy hauling. Oxen were said to be too slow, but this was vigorously denied by some freight haulers who claimed that ox teams could be trained to travel fast. While ordinary ox teams made about fifteen miles per day hitched to covered wagons or freighters, there were "lightning" teams that traveled twenty-five miles a day with heavy loads. Usually mule teams were considered the fastest freight haulers in provisioning the West, but there were ox teams which were both tougher and faster.

The high point in the use of ox teams came around 1850-60 when a firm known as Russell, Majors and Waddell held government contracts to supply the military and civilian operations in the opening of the West. Buffalo Bill Cody and Wild Bill Hickock were both said to have gotten their start as bullwhackers and muleskinners with this outfit. In 1858 Majors reported that he employed 3,500 wagons, 40,000 oxen, 1,000 mules, and 4,000 men —giving an idea of the relative

THIS COW AND CALF, along with the champion steer shown on the previous page, is one of the superb engravings published in Prairie Farmer in an 1886 issue. The cow is identified as Wild-eyes Winsome 4th and calf, owned by H. F. Brown of Minneapolis.

DURHAMS, both horned and polled, are usually regarded as an off-shoot of the Shorthorn, but these cattle have persisted as a separate breed in some places. The upper woodcut of the bull Archer was carried in an 1853 issue of American Agriculturist, indicating very early origin, almost simultaneous with the traditional Shorthorns. The lower engraving is of three Polled Durhams, Mollie Gwynne, Nellie Gwynne, and King of King, owned by W. S. Miller of Ottawa County, Ohio. An American Polled Durham Breeders Association was organized in 1889, and a herdbook was launched soon after. The polled characteristic was emphasized in registering animals.

The property of Col. J. M. Sherwood, of Auburn, N. Y.

HEREFORDS began to nudge the Shorthorns out of first place on the American beef scene as the 1800s drew toward a close. The competition was fierce and the arguments were heated. Here are two superior animals from the 1880s that helped bring about the change. Engravings are from the Orange Judd Farmer of 1886, probably borrowed from other papers. The bull is Anxiety 2nd, owned by G. W. Henry of Ashkum, Illinois. The cow is an import named Venus, owned by Adams Earle, Lafayette, Indiana.

BELOW is a really remarkable engraving of a group of Herefords on the Cosgrove farm at Le Sueur, Minnesota.

value of the ox and the mule in this huge operation. At peak this outfit employed 75,000 oxen and 5,250 wagons.

In spite of the superior record of oxen as draft animals, the shift to horses picked up momentum during and immediately after the Civil War. Railroads and steamboat lines took over much of the heavy, long distance hauling. The invention of the steel plow to replace the old wood and cast-iron sod breakers reduced by about a third the power needed to break the virgin prairie. Heavier horses were being bred which were more capable of handling heavy farm work. Horses had the drawback of needing more care and feed grain, but they were destined to take over for more than a

half century until they in turn were threatened by the gas tractor.

BISON WAS ALREADY THERE

When the pioneers came west, they found the Great Grass Plains already in full possession of a rugged member of the *bovinae*, namely the American bison, which had moved into North America on the land bridge to the north and found the western plains increasingly to their liking. There were millions of them, thoroughly at home in the ecology of the West. They were a chief source of food for the native Americans, the Indians, and also sustained the trappers and earliest explorers. It wasn't their meat that spelled their doom. It

Star Grove 9th 17429.
Countess Clive 14223.

Wild Eyes 11559.
Eolah 28810.

Miss Poppy 16936.
Bonnyface 21605.

was the popularity of buffalo hides in the East and in Europe.

Cattle took over the plains and the bison nearly disappeared, only to be saved by conservationists and hobbyists. The deliberate crossing of cattle and bison was attempted early by ranchers, with only moderate success. Today, spurred by the call for a meat animal that can be fattened without grain, there is new interest in cross breeding the two.

While the bison was very much in evidence in North America before the discovery of the new continent, there were no cattle here until Columbus landed an Andalusian bull and a group of heifers on Hispaniola Island in the Caribbean Sea in 1493.

EVEN BEFORE COLUMBUS

Correction! There were. The Vikings colonized Greenland well before the year 1000 and brought with them red and white Norse cattle, along with sheep, goats, and horses. In 1004 the Vikings extended their colonization to Vinland after an earlier discovery of the mainland by Leif Ericson. They brought livestock, according to accounts in the Norse sagas which are considered authoritative. The colony lasted only a few years. Some settlers lost their lives in battles with the natives and others left to return home to Iceland or Greenland. As far as is known the livestock were completely wiped out.

The Greenland colonies survived for several hundred years, taking advantage of a warm cycle in the climate. There is evidence that cattle thrived for a time, but as the weather turned cold again, the livestock perished along with the people. Some sort of plague such as the Black Death may have been the cause, rather than the weather.

In 1598 the Marquis de la Roche planted a French colony on an island off Nova Scotia and found wild cattle and sheep. It was deduced that these animals had come off shipwrecked Spanish vessels that had come to their end along this coast. They were not the remains of the Viking colony on Vinland.

So it seems Columbus was the first to introduce cattle into the Western Hemisphere that had the stamina to survive on their own. It has been noted that Lower Andalusia in Spain had just the climate—dry, hot, and rugged—to fit the early Spanish cattle for ordeals in Mexico and the Southwest.

Even as the indestructible bronco—mustang, cayuse, or whatever—took possession of the West and greeted the pioneers as they arrived, so also another indestructible maverick made his mark and became a legend, the Texas Longhorn.

SPANISH WERE STOCKMEN

The Spanish who at one time occupied Mexico plus all our southern tier of states from Louisiana to California, were good livestock men. Their stock was also tough and fit for survival on its own, which may be due at least in part to the similarity between the climate in our dry southern states and the climate in Spain. At first, the transporting of stock was a military matter, but soon the padres took over as the church became active and established missions all the way across these states. The padres systematically stocked these stations with horses, cattle, sheep, and goats. Many of the animals escaped into the wild. Others were abandoned when Apaches raided the ranches and ranchers were forced to draw back to safer areas.

The significant thing is that horses and cattle thrived and multi-

ABERDEEN-ANGUS cattle were later arrivals on the beef scene in both Britain and America, but they lost no time once they got started. Toward the close of the 1800s they began winning at shows and reached a peak in the 1930s and 1940s, capitalizing on a smallish, blocky body of high quality. The animal below was the imported steer Benholm, winner at the Chicago International in 1885. He was owned by railway magnate Jim Hill of St. Paul, founder and president of the Great Northern Railway.

Jim Hill was a great booster for good livestock, but it is said that in his presidential office there hung a picture of a disreputable looking scrub cow. When people asked why he displayed such an animal instead of his own prize winning livestock, he replied: "This is the only scrub cow ever run over by a Great Northern train. It cost us only $37. All the rest were valuable blooded stock."

75

RED POLLED cattle were widely dispersed throughout America in the 1800s, being regarded as a dual purpose breed. On this page are types showing considerable variation. The lower panel is an engraving of the imported cow Ocean Maid No. 96, an excellent dairy type, shown in Prairie Farmer, 1888. The upper panel carried in Country Gentleman in 1877 and described as Norfolk Red Polled cattle, shows rather spindly animals.

BELOW are beefier Red Polls, probably more typical of the general run of the breed in this country.

plied on their own. By the time Texas became a republic and most of the area, including California, was annexed to the United States, the whole region was swarming with wild cattle of a breed that beggars description. I cannot begin to take the space and time to describe them here. Everyone interested in American history should read J. Frank Dobie's *The Longhorns* (copyrighted in 1941 and available in several editions, including paperback). While these fabulous, indestructible cattle are known mostly as Texas Longhorns, they and their close relatives covered the whole Southwest. The territory between the Nueces River and the Rio Grande was considered the "nursery" of this wild and wonderful breed of cattle. They were wilder and smarter than the grizzly or the buffalo.

THEY PUT UP A FIGHT

When settlers worked their way across the Southwest establishing small farms and ranches, they regarded the Longhorns as indigenous beasts and called them mustang cattle, Spanish cattle, or just plain wild cattle. These animals had to be dealt with because they were constantly stealing stock from the tame herds, spiriting them into the mesquite jungle, and generally making a nuisance of themselves.

Eventually many were branded and the bull calves castrated so that after a fashion they were ranch cattle, partly mixed with English cattle from the East. These semi-wild ranch cattle were the ones that made history up and down the West, tramping up the Chisholm Trail and other less known thorough-

fares on their way to markets a long way off.

It is said that more than ten million head were trailed out of Texas, first toward the East, Louisiana, Missouri, and as far as the slaughtering plants of Cincinnati and Chicago; then to the rail heads, most famous of which was Abilene, Kansas. Later they were trailed north to take over the grassy plains of Nebraska, Wyoming, Montana, and the Dakotas. Even these northern ranges became overstocked. It remained for fierce blizzards and long winters to starve out the cattle and cut the herds down to size.

FARMERS LIKED TAMER STOCK

While the Longhorn and all its tough shirt-tail relatives were conquering the West and making available large quantities of hides and tallow to be sent East just before, during, and after the Civil War, the same West was being settled by farmers and ranchers who opted for the tamer and blockier English breeds of cattle. Whereas the Longhorn could not be tamed by weather, wild beast, or even cowboy, it was eventually subdued by cross breeding. In fact this process continued for so long that the Longhorn, like the buffalo, almost disappeared.

In 1926 the U.S. Army rounded up some of these mustang cattle and tried to resettle them in a forest

AT LEFT

BACK IN 1887 the cattle we call Brown Swiss today were still referred to in American Agriculturist as Brown Schwytzers. They were regarded as a new breed in America, but they were given high praise as one of the purest dairy breeds in Europe. They were compared to the Channel Island cattle for production of high butterfat milk. The engravings shown here indicate a conformation much like that found in today's breed in this country.

PRAIRIE FARMER

preserve. Probably the principal motive for preservation was the need for authentic herds to use as props in western movies. If you are of an older generation and knew something about ranch cattle, you were no doubt irked during the Thirties and Forties by high bred Herefords being driven up the Chisholm Trail in an otherwise quite authentic cinema.

In an effort to restore the Longhorns, hobby breeders scoured Mexico for stock and did some back breeding (not unlike the zookeeper in Germany who tried to restore the Aurochs). Now we do have a few herds of fairly authentic Longhorns. In 1974 when public opinion cried shame over all the corn and soybeans (people food) going into finished prime cattle in the feedlots, somebody discovered that Longhorn meat is surprisingly tender, and Longhorns can make fast gains on grass alone. This may be just press propaganda—but who knows! One thing is sure, a resurrected Longhorn won't be the same.

NORTHERN EUROPEAN CATTLE WIN

It's time to bring this story back to the Northern European cattle that crossed the Atlantic with the first colonists, stimulated the interests of the importers and breeders, and eventually took over this country. They too have a varied and romantic history.

Yes, there was an English Longhorn breed too, but there is no known connection with the Texas breed. These particular Longhorns are believed to have been derived in England from a mixture of Flemish and Normandy cattle that were imported into England after the Norman conquest in 1066. They are related to the earliest Shorthorns and Herefords, but did not retain their popularity long enough to become a factor in the building of the American herds.

Concern for cattle breeding in England goes back at least to the reign of Henry VIII when laws were passed prohibiting the running free of stallions, bulls, or rams that did not live up to certain specifications. This could be a precursor of the practice in America of licensing stallions before they could be kept at stud.

In the 1700s quite a variety of cattle breeds were distinguishable in Britain, the hardy longhaired mountain breeds, Scotch and Welsh Kyloes; Alderneys or French cattle; polled Galloways, Longhorns, Herefordshire Browns, Shorthorns, and Dutch cattle which are related to modern Holsteins.

It was to be expected that the European breeds—if they can be called that so early in the game—would filter across the Atlantic into the American colonies. Earliest colonists from England seem to have favored the red Devons. The Danes brought their yellow cattle that were

especially good as oxen. The Dutch brought their black and whites. Division into milk and beef breeds seems to have come later. After all, any breed was expected to do fairly well in the milk bucket in those days.

Cattle raising was one of the first successful businesses in the colonies. Export of hides and tallow began long before the Revolution. It seems odd today that tallow was a more valuable commercial product than meat. There were many uses of tallow in pioneer times, and preservation was not a problem as with meat.

BIG ANIMALS WERE POPULAR

This accounts for the great interest which prevailed until the present century in huge cattle that yielded a lot of tallow. Four- and five-year-old steers commanded a good market. The shows stressed huge animals. Steers weighing 3,500 pounds were not uncommon and attracted great attention.

Improvement of cattle became a passion in both England and America before much attention was paid to horses.

Separation of cattle into beef breeds and milk breeds was a slow process. In colonial times most cattle were dual purpose—one might call them triple purpose because draft cattle were still an important factor in farming and in freighting. They still are in Central America and

BEN JONSON, THE PRIZE HOLSTEIN-FRIESIAN STEER.

in many parts of the world.

The so-called dual purpose breeds remained in vogue for a long time. Milking Shorthorns hung in there well into the 1900s. Red Polls (Suffolks), Danish Reds, Devons, Brown Swiss, and even Holstein-Friesians held their popularity as dual purpose breeds until very recently. Holstein steers turned up in the shows back in the late 1800s. There are still feeders who swear by them. The present vogue for rangier, faster gaining, red meat cattle that finish well on grass may turn up all kinds of unusual crosses reminiscent of the dual purpose types.

However, our present "fine tuned" dairy breeds and the specialized nature of modern dairying make it unlikely that intermediate cattle will ever be kept for milk again. Meat cattle will have to build their popularity on rapid gains, plenty of milk to get a calf off to a fast start, and an abundant yield of red meat, not necessarily marbled with fat.

AT LEFT

HOLSTEIN STEERS, every once in a while, are pushed hard as beef animals. This usually occurs when the market is down for the fancier beef and it appears sensible to go for plainer cattle that can be fattened on cheaper feed. One such rise in the popularity of Holstein steers came in 1891 when the steer in the upper illustration, named Ben Jonson, was awarded a first prize at the Chicago International. He was owned by M. L. Sweet of Grand Rapids, Michigan.

Holsteins were long known as the big black and white Dutch cattle. They were grown for beef as well as milk in Europe and still are to this day. The Low Countries went in for large livestock in a big way, both horses and cattle. The engraving at left was published in Prairie Farmer in 1888, evidently to show how massive were the Holland-Flemish-Netherlands cattle that were the forerunners of the Holstein-Friesian.

HOLLAND, FLEMISH, OR NETHERLANDS COW

Since the period covered by this book is roughly 1840 through 1920, it is well to remember that the marketing of dairy products was an entirely different proposition from what it is now. During the 1800s the products in the market were butter and cheese, turned out to a large extent by women in home dairies. The word creamery was first applied to the dairy room on the farm. The creamery as a marketplace, with its assemblage of farm teams bringing in cream, with a trained butter-maker in charge, built up slowly during the 1800s. The miscellaneous cows in the farm herd did pretty well in furnishing cream for this business, when you discount the laborious work of extracting the milk by hand from "critters" with udders hairy or smooth, teats large and small. The "cash money" from butter and cheese sold at general stores or from house to house was important in buying the necessities that had to be bought at the store.

The dairy breeds of today are pretty much the same as those that took over from the dual purpose cattle around the turn of the century. The Holstein - Friesian, Guernsey, Jersey, and Ayrshire cattle were the primary milk breeds then and still are. Belted Dutch had a brief popularity, but are now little known. Brown Swiss built their popularity later.

THE DAIRY BREEDS

Holstein-Friesian. While this dairy breed was originally known as Dutch and arrived on American shores early, it has become a nearly universal dairy breed. American Holsteins stand pretty much on their own feet. They have not depended on importations to improve the stock for a long time. Through the years there have been many refinements in type and markings, but the breed stands very much on its record of production of quantity milk, with less emphasis on butter-fat content than in the other dairy

breeds. Holsteins have been the chief beneficiary of the trend to stress the protein rather than the fat content of milk. They are undoubtedly the most numerous dairy breed in America, and probably in the whole world.

Guernsey and Jersey. These are often called the Channel Island breeds, and have been mistakenly called Alderneys. Because of their origin on islands, it was possible to keep the breeds purer. There was in fact legislation in the Channel Islands to keep out foreign cattle as early as the 1700s. Both Jerseys and Guernseys are smaller cattle than Holsteins and traditionally give much richer milk. In later years the Guernseys were bred larger, probably to meet Holstein competition, but the yellow color of the milk, due to high carotene (vitamin A) content, was carefully retained by the breeders. Guernseys are mostly spotted light brown with white. Jerseys are noted for their beautiful heads and fine-bone bodies. They have a characteristic fawn color with shadings of black which give a striking appearance. Some engravings of early Jerseys show them as spotted like their close relatives the Guernseys.

The arguments over the relative value of the butter breeds and the

THE CARTOON above will recall to many a former farm boy the pestiferous task of teaching a calf how to drink milk out of a pail.

milk breeds go way back. In a hot exchange between a Holstein man and a Jersey man, the latter was commenting on the thin, blue milk of the Holstein. He said, "If you milk a Holstein cow into a pail and then drop in a silver dollar, you can see the dollar right through the whole pail of milk!"

Retorted the Holstein man, "If you milk a Jersey cow into a pail and then drop in a dollar, you'll find the milk won't even cover the dollar!"

So the battle as raged through the generations. The cholesterol scare has resolved the argument in favor of the Holstein—at least for the time being.

Ayrshire. The Ayrshire is a Scottish breed which once leaned toward dual purpose, but which has gained increasing popularity in the dairy field, especially in Canada, some parts of Europe, and New Zealand. It is believed to have been derived from the Shorthorn and the Channel Island cattle. The breed is noted for its heavy production of milk and for its ability to make do with forages.

HOLSTEIN-FRIESIANS *are today the most popular dairy breed in America and in Europe, and they have held that position for a long time. Here is an outstanding pair owned back in the 1880s by Smiths, Powell and Lamb, Syracuse, New York. The cow was Lady Fay 4470 and the bull Netherland Prince. These are listed as imported, but except for the first good animals, America has produced its own pedigreed stock without infusions of European blood.*

BELOW is a group of Holstein cows owned in 1888 by Wilcox and Liggett of Benson, Minnesota. At right, for the purposes of type comparison, is a more modern Holstein sire, Man O' War Ormsby Posch, champion of the Waterloo Dairy Congress in 1936 and owned by Maytag Farms. Most desirable type among Holsteins has not varied greatly.

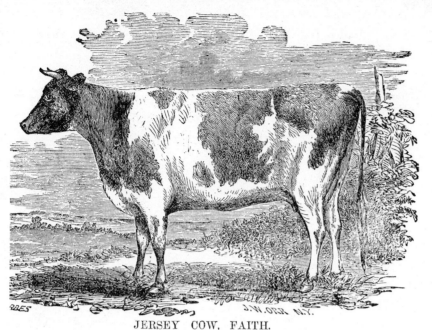

JERSEY COW, FAITH.

Calved 1850. Imported August, 1854, from the Island of Jersey, by J. A. Taintor, for J. Howard McHenry.

THE BEEF BREEDS

Three principal beef breeds have occupied the limelight in America for most of the years with which we are concerned here. They are the Shorthorn, Hereford, and Aberdeen Angus. At the recent end of a century of cattle breeding history in America, we have any number of crosses which have involved these breeds. For instance there is the Santa Gertrudis which is mostly a cross of Shorthorn and Brahman. Recent years have also seen the revival of at least two old French breeds, Charolaise and Limousin, and a Swiss breed, Simmenthal, to use alone and in crosses; plus at least a half dozen other breeds and strains which have received less attention.

In the latter part of the 1800s we had a cluster of beef breeds, all from Europe, that enjoyed limited popularity in America, but they were eventually crowded out by the big three. Among them are Devon, Red Poll (an amalgam of Suffolk and Norfolk), Sussex, Galloway, Highland, Dexter, Kerry, English Longhorn, and others. Any list would leave out some breed which is considered important by someone.

Here is a quick look at the big three:

Shorthorns. First identifiable Shorthorns, also called Durhams, came to America in 1791. The breed was put together with careful crossing and selection by Charles and Robert Colling around 1780 from

OPPOSITE PAGE

THE JERSEY BULL shown here is Litchfield [674] who won the Centennial Award given by the American Jersey Cattle Club in 1876, just 100 years ago. The owner was F. Ratchford Starr, Litchfield, Connecticut. The cow is a famous Jersey female of some years later, Lady Mell 2nd 1795, owned by Charles F. Mills, Springfield, Illinois.

ABOVE

JERSEYS AND GUERNSEYS, known as Channel Islands breeds, look more and more alike as one goes back into history. This rare woodcut of a Jersey cow calved in 1850 was printed in 1858 in Northwestern Farmer, published at Dubuque, Iowa. Markings are that of a Guernsey today. Also shown here is the young bull Diavolo, an engraving taken from the Rural New Yorker of the 1880s.

85

some very outstanding individual cattle which had been bred by a whole series of improvers. At any rate, the Shorthorn was well on its way by the end of the 1700s and enjoyed great popularity in England, Scotland, and America. It was without question the top breed in America until around 1885, when it encountered serious competition from both Herefords and Aberdeen Angus. Shorthorns, though somewhat eclipsed in recent years, have enjoyed nearly world domination among beef breeds. They are adaptable to different climatic and feeding conditions, and do exceptionally well when crossed with other breeds. There are many strains, large and small, leaning toward milk or toward beef. They come all red, all white, or different shades of red roan in between.

Shorthorns were great favorites of importers and breeders in America during the 1800s. They dominated the big shows and did a great deal to build the exalted purebred tradition in this country.

AT LEFT

HERE ARE two Guernsey cows from the 1880s, showing a type not too different from today. Top engraving is of Bonnie Lassie at two years of age. Lower illustration is of an imported cow, Florence of Guernsey. Perhaps they can be traced through the herdbooks.

CLAIM TO BE OLDEST

Herefords. As an identifiable breed the Hereford, or Herefordshire, is claimed to be even older than the Shorthorn. The origins of the breed are traced to a Tomkins family of breeders in Herefordshire, England, around 1742. This family dominated the breeding work well into the 1800s. The Tomkins breeders did not try to stabilize the Hereford markings as red with white face. Their cattle were sometimes grouped in four classes, mottle-faced, light gray, dark gray, and red with white face. Today's white face was achieved later by careful line breeding. Herefords were introduced into the United States in 1817 by Henry Clay. They made rather slow progress in this county until around 1870 when T. L. Miller of Illinois, acknowledged by many to be the father of the Hereford tradition in this country, took over and pushed the breed to outstanding success. Herefords adapted well to life on the range and soon became recognized as the outstanding range breed. They

DUTCH BELTED was a very fine-boned dairy breed that made a start in America but never gained much momentum. This is the cow Lady Aldine, illustrated in the Scientific American as a part of a series on farm livestock run in the 1880s.

GUERNSEY COW Rosebud 1037, shown below with calf, doesn't seem to carry the Guernsey type, but that can be the fault of the engraver. Guernseys tend to look more like Jerseys, and vice versa, as you go back in years.

87

Vanzant Del.

PRAIRIE FARMER

have been subject to the usual ups and downs of type among breeders and fanciers. For a time the most favored Herefords were small and blocky like the Angus. But a reaction set in based on the theory that Herefords would lost their excellent performance as range cattle if they were turned into "toys of the show ring." Now a larger and more rugged animal is in vogue.

Aberdeen-Angus. This Scottish breed is black, hornless, smaller than most beef breeds, but very blocky. The quality of its beef has long been considered the best of any breed. For many years in both Britain and America the Angus won more than its share of the big prizes because judges favored a relatively small package of very good beef.

The Galloway breed undoubtedly had a part in building the Angus, which were known as "Doddies." The Angus was not introduced into America until 1873 after which it enjoyed rapid growth in popularity. The herdbook was started in Scotland in 1862, sharing for several years its records with the Galloway. The Angus, like the Shorthorn, has been favored in beef-beef and beef-dairy crosses. It manages to hold its own rather well in this country, after slipping somewhat from the peak of its show winnings which occurred in the 1940s.

WHIMS OF THE JUDGES

Much could be said of the influence, both good and bad, on the beef breeds of the big shows, most especially Chicago's International. I have no doubt that the judges tended to run the beef breeds into the ground for a time with their preference for small, blocky, highly finished animals. But the tide reversed itself, the judges began to go for the rangier crossbreds and the rangier specimens of the pure breeds.

The beef breeds have been much influenced by imports down to comparatively recent times.

AYRSHIRES *are a Scottish breed noted for their ability to produce a great deal of milk on sparse pasturage. The woodcut of the bull at the top was published in the Cultivator of 1852, which accounts for the rather crude illustration. This is the bull Dandy owned by J. C. Tiffany of New York State and a winner in eastern shows of 1850 and 1851, indicating that the breed got an early start in America. The cows are illustrations from magazines of the 1880s. The lower engraving shows type very much like that accepted today.*

AT RIGHT

DEVONS *are believed to be among the first dual purpose cattle brought to America from England. Along with the Red Polls, they account for the great predominance of the color red in early American cattle. The top engraving of the bull Baltimore comes from a Country Gentlemen issue of 1856, which explains the crude illustration, probably a woodcut. The bull won top prizes in New York State as early as 1847. He was bred by H. N. Washbon and owned by J. W. Collins, both New York importers. The breeding is listed in Davy's Herdbook. Lower illustration is of the bull Carlos 2013, owned in the 1880s by J. W. Morse and Son, Verona, Wisconsin.*

ABOVE

CHAROLAISE cattle from France have become tremendously popular in America since 1950, but don't get the impression that they are a new discovery. They have been around a long time, as attested to by these sketches published in Prairie Farmer in the 1870s. There is no record of their being actually imported at the time, but they were known and thought worthy of notice. These are described as a typical cow and steer.

PRAIRIE FARMER

AT RIGHT

THE HAIRY, HARDY Scottish breeds of cattle never went to the top of the class in America but they have their followers even today. The black bull, Harden, a prize winner in Britain, in the top engraving is a Galloway, a breed which enjoyed a brief popularity this side of the Atlantic. The Highland cattle shown below are considered something of a novelty in this country but their extreme hardiness led to their being tried out in western mountain country. They are now being used in crosses in the search for really hardy and thrifty grass cattle.

SHROPSHIRE was the name given this breed of cattle in England going back well beyond the 1700s. This woodcut is taken from an issue of The Cultivator of the year 1852. The breed goes back much farther. The livestock historian Youatt credits this breed with being a forerunner of the Hereford.

WELSH cattle, of which this cow is believed to be typical, are among the oldest in Britain. They are said to have come from the same stock as the Wild White Chillinghams. Coming in practically every color, they appeared to be high on the rump, flat sided, with the greatest depth in front. Horns were a rich yellow, tipped with black. While fairly successful as a localized breed, they have not been generally popular.

KERRY cattle originated in Ireland but have raised interest from time to time among breeders. They are small cattle, low and long, with a good girth. They are very hardy, well suited to thin and rocky pastures and raw windy weather found in their native haunts. They give very rich milk, 10 to 15 test according to some reports. The preferred color is black with some white along the spine.

WHAT KIND of cattle did they have Down Under? Today there are mostly British breeds, although New Zealand has been perfecting new dairy strains that are really distinctive. In its attempt to serve its readers during the 1880s Prairie Farmer came up with sketches of many little known breeds from all parts of the world. Among the sketches was this one of the Australian champion dairy cow Violet. It is hard to associate this animal with any particular breed, but she was said to be a Milking Durham, nearest thing to our Milking Shorthorn.

ENGLISH county names keep turning up on all kinds of exotic livestock. Local names must have persisted for a long time, even when overshadowed by the most popular breeds in the national shows. This is a Sussex heifer, evidently considered worthy of a place in Prairie Farmer's list of interesting cattle.

BRITTANNY cattle were spread over five departments of France. There were once large numbers of these cattle, but they have not spread widely beyond their own area. This may be due to their small size, grown cows weighing only about 400 pounds. The breed has ancient origins. Colors are brown to black with patches of white.

THE LIMBOURG breed of German cattle actually had its origin in the Belgian province of that name. These cattle were also called Wurtemberg. The cows weighed 700 to 800 pounds and mature oxen from 1,600 to 1,700, indicating that they were good general purpose animals. The color was reported to be silvery yellow.

A PRAIRIE FARMER writer in 1889 noted that the Kingdom of Baden, Germany, had four distinct breeds of cattle, Hinterwalder, Messkircher, Oldenwalder, and Necker. The cow shown here is a Messkircher. While distinct lines to local people, these breeds bore considerable resemblance to each other. Americans visiting Germany today will encounter many of these same cattle, which look like something between a Guernsey and a Brown Swiss.

THIS IS a Voigtland or Saxon cow, considered a separate breed about 100 years ago. The German breeds are practically all dual purpose, kept for milk and meat. In years past they also had to work in the fields, a task that fell to both cows and bulls.

SALERS was another breed of French cattle of which this cow and calf are representatives. They are mountain cattle from the district of Mauriac. They were very good work cattle and also fattened well. Solid red like the British Devons, they may be related somewhere back in history.

BELOW

LITTLE IS KNOWN about cattle raised during this period in Russia and the Slavic states. This white native Hungarian bull was selected by Prairie farmer as a representative.

ABOVE

ORANGE JUDD FARMER, in its August 31, 1895, issue, devoted its front page to the wonderful Simmenthal cattle from Switzerland. A herd of six bulls and eight cows had just been imported by Theodore A. Havemeyer of New Jersey. He proposed to cross them with Jerseys. Evidently this importer was a bit ahead of his time because very little was heard of this breed until about 1950 when they became popular in cattle breeding experiments in this country.

WALLACES' FARMER AND DAIRYMAN
A WEEKLY JOURNAL FOR WESTERN FARMERS

VOL. XXII.   DES MOINES, IOWA, FRIDAY, JAN. 29, 1897.   **NO. 5.**

The Proper Covering for Grass Seed.

It is with some surprise that we see the question raised in agricultural papers as to whether grass seed should be sown or planted in the sense of being sown on the surface or covered like other grains, and the experience of practical farmers requested as to which of these methods is best. For advanced Iowa farmers this question became general that grass seed would grow in any part of the state. Our old readers do not need any information or instruction on this subject. We have endeavored to indoctrinate them thoroughly with the idea that grass seeds in any country must be sown at a depth which will give them moisture, heat, and light, and we might add air. What that depth should be depends on the soil and the season. If they are the same covering unless the soil be a heavy clay. Even in this case the covering may be the same, as lighter covering is needed in these kinds of soils for any kind of grains. In seasons of frequent rainfall at the time of sowing, damage might ensue if grass seeds were put as deep as the cultivator covers the grain, and in this case we would sow and harrow. Under any circumstances we would give them the seed should be covered a sufficient depth to insure the moisture necessary to make a prompt start and vigorous growth. The farmer is the best judge as to how this is to be done. Our old readers do not need this instruction, but so many new subscribers are being added daily to our lists that we think it necessary at this time of year to call attention to what may be an old and threadbare story to those who

Royal Tecumseh 16535 S. Owned by C. D. Luther, Marcus, Iowa.

has been settled for some time. Many thousands of dollars worth of grass seed was thrown away ten or fifteen years ago by following the Eastern method of sowing it on the surface of the ground just after the harrow and trusting to the settling of the land to cover it. To this more than anything else was due the impression, which it took a long time to correct, that the tame grasses would not grow even in eastern Iowa. As settlers moved West the same impression prevailed and it was only after they learned from their own experience how to sow or harrow and how to cover that the conviction placed too deep, sufficient light can not reach them and they will not germinate no matter what the moisture, heat, or air. If they are placed on the surface where moisture is lacking, heat, light, and air will do them no good, and some seeds will lie in this condition for years without growth. This is especially true of clover seed, which has a thick covering and must be soaked up before it will grow under any circumstances. In dry seasons the grass seed should be put as deep as the farmer would cover oats or spring wheat, and he need not hesitate to sow with his spring grains and give them covering that a smoothing harrow gives.

In dry seasons the depth of covering is not all that is important. The soil should be compressed around the seeds in order to secure germination. When a boy covering corn with the hoe, if the soil was dry and loose we always stepped on the hill in passing, the object being to compress the soil around the grains of corn and insure rapid growth. If it was wet, this would be an injury rather than an advantage. In short, therefore, the important point is to get this principle clearly before the mind, that grass have been reading after us for a number of years.

Sugar Beet Growing.

Elsewhere we print the conclusion of a very lengthy and exhaustive paper read by Professor Albert Myers before the Iowa Agricultural Society at its recent meeting in Des Moines. The subject is one that the farmer would do well to turn over in his mind. The time is coming when we will cease to pay a hundred million gold dollars to foreign countries annually for sugar. It can and will be grown at home in due time to the profit of the farmer and to the benefit of the general public.

SWINE

OPPOSITE PAGE

WALLACES FARMER got an early start in its superb coverage of corn and hogs. This front page from an 1897 issue features a prize Poland China boar, undoubtedly stylized by the artist with an eye to showing what a lard hog ought to be. The lean bacon hog was to come nearly a half century later.

"Why have some animals submitted readily to domestication while others have remained wild? Why should the dog be man's loyal companion while the closely related fox and wolf remain enemies? Why did the horse remain a tractable and co-operative servant when the zebra is still savage? Why did the ox earn an important place in our economy when the bison suffered near extinction rather than mingle with men? Obviously there is something present in the temperament and disposition of certain animals that helped build civilization, which is lacking in others. One may well speculate why the wild boar, *Sus scrofa*, became a bulwark of our agricultural production when his near relatives—peccaries, wart hogs, African forest hogs, and the babirussa—have successfully withstood all intercourse with humans."

The above quotation is from *PIGS from Cave to Corn Belt*, a very good history of swine by Towne and Wentworth (see bibliography for further documentation). The lowly hog did not carry knights into battle as did the horse, nor haul the heavy loads of growing civilizations, as did the ox; but century in and century out, he has probably fed more people more efficiently, in more parts of the world, than any of our other domestic animals. Fresh killed, salted, pickled, smoked, and more recently frozen, pork has been pretty much the poor man's meat all over the world.

The hog was long considered the most efficient of the converters of feed into meat and fat. Recently domesticated poultry has challenged that record, but the hog is probably the most versatile of the meat producers. In this country, our great abundance of concentrate grains, projected into computerized rations and automatic feeding, has caused us to lose sight of the hog's versatility as a feeder. The old saying that a hog will eat anything is very close to true. The more primitive appetites of the pig may seem far removed from those displayed in the automated feedlot. However, if modern pigs are turned loose in the wild, they adopt feral characteristics very rap-

97

idly and become really wild in a generation or two. They will eat and thrive on seeds, roots, fruits, nuts, mushrooms, snakes, larvae, worms, eggs, carrion, mice, and other small animals. This does not mean they do not like some things better than others. Farmers harking back to the introduction of hybrid corn will remember the hog's preference for the open-pollinated varieties.

At one time "grazers" among hog types were considered most desirable because they could do well on cheaper feed and especially cheaper protein. It seems likely that as concentrates become more expensive and more likely to be considered "people food," we may return to breeding hogs that can use forage crops efficiently as well as grains, and vegetable proteins as well as animal proteins such as tankage.

AT LEFT

SWINE OF TODAY descended from some pretty rough characters which had handled themselves successfully in the wilds for millions of years. Top illustration is of the wild boar, Sus scrofa, from which our domestic swine are derived. The middle illustration from an old farm magazine is of an early domesticated English hog, not too far removed from the wild boar. The lower drawing is of Chinese swine which turn up frequently in histories of early hogs, indicating that today's swine are of both European and Asiatic origin. These were exhibited at a show of the New York Agricultural Society in 1851 by John Delafield of Geneva, New York. It is not recorded whether the animals arrived by land or sea, or whether they were a European breed still using the Chinese name. The brief item associated with the illustration reports that China had several breeds of swine around 1850, that they were noted for fattening extremely well and that they produced twelve to fifteen pigs per litter.

SWINE ARE OLDEST OF ALL

Swine are without doubt the oldest of all the domestic animals. They are believed to have been around for forty million years and have undergone fewer physical changes than almost any other mammal. Perhaps this has been due to the variety of foods a hog can handle, and to its relatively modest size as compared with the evolutionary giants that failed to survive. Before the Ice Ages *Sus* was well scattered over the earth and doing very well by itself. Its fearlessness, its ability to defend itself, and its no-nonsense habit of tending to its own affairs made it a strong competitor and indefatigable survivor.

The Ice Ages gave *Sus* a hard time because this otherwise hardy animal had very little protection in its bristles against cold and very little protection against heat without sweat glands to furnish a cooling system. The hog has always been a cool forest animal, a fact which is evident even today. Today's high-bred sows will take to the woods to have their litters if given a chance. And when domesticated pigs become feral (return to wild state) they choose the woods as their favorite habitat and quickly change their habits to night feeding.

In pioneer times hogs were kept half-wild simply by virtue of the fact that sows were permitted to farrow in the woods and run wild until fall when they were with some difficulty enticed back into captivity and fattened on corn. It was not uncommon to hear complaints that cornfed hogs did not have the flavor of those that fattened themselves on acorns and mast. The much touted southern cured hams earned their reputation not entirely from the curing but also from the way the hogs were fed.

Even before the Ice Ages swine had occupied every continent on earth except Australia. Although the glaciers and their vicious cold cut down on the number of branches of the species and the extent of the

HOGS WILL GROW to enormous weights, as indicated by this advertisement in Nebraska Farmer at the turn of the century. We also learn that the right feed was believed to cure or prevent hog cholera, a theory which must have been knocked in the head soon after.

The lower advertisement from the year 1888 shows that the engineering of hog houses to save newborn pigs was undertaken a long, long time ago. This one featured an inclined floor and "pig porch retreat," complete with raisable doors to expose the little pigs to fresh air and sunshine. There is no clue to the inside arrangement.

range, some member of the pig family managed to survive and flourish almost everywhere where there were moderate temperatures and some kind of forest cover.

However, the swine that arrived at economic importance in the affairs of man belonged to the *Sus scrofa* branch of the family. They were situated in Europe and Asia, where they were domesticated just about the time when Neolithic Man, between 7000 and 3000 B.C., began to change from a wandering nomad to a farmer settled in one place. This was the signal for the domestication of the wild boar, coming some time after horses, cattle, and sheep had joined forces with human beings. Long before that, however, the wild pig had been an important factor in the meat diet of early man.

MENTIONED IN LITERATURE

References to domesticated swine turn up in Egyptian literature as early as 2900 B.C., in the Greek classics such as the *Iliad* around 1000 B.C., and in Britain soon after that. However, there is reason to believe that the Chinese regarded swine as property as early as 4900 B.C. Whether our swine today were brought from the East by the migration of the Neolithic people, or whether they were tamed from the European wild hogs is hard to determine. There was a Chinese hog among the first strains that were subjected to breeding procedures in Europe. Even today we have the Poland-China which is neither very Polish nor very Chinese.

The two greatest swine breeding nations today are China and the United States. China, with its enormous task of feeding 900 million people, is said to raise 120 million swine annually, while the United States ranks a poor second with around 75 million. This tends to discredit some of the current "wisdom" about the necessity of populous nations abandoning livestock in favor of a cheaper vegetarian diet.

TOOK OVER NEW COUNTRY

The swine that were brought to America with the first discoverers and settlers were, of course, European in origin, and nondescript in character. Nevertheless they were tough, prolific, and quite capable of "taking over" new country, with or without the aid of their masters.

Columbus brought over eight porkers on his second voyage in 1493. The hogs prospered even though the human beings had more or less trouble getting established. They multiplied and ran wild, taking over the canebrakes and jungles of the islands in a very few decades. No doubt the Spanish expeditions to the mainland brought more breeding stock which escaped into the wild. By the time the British began to colonize the east coast, from Virginia to New England, hogs gone wild were scattered all over Mexico and what are now our southern states—all the way to and including California.

From then on the line between wild and tame pigs was not a very clear one. The settlers found it both cheap and convenient to let pigs fend for themselves in the woods. They made no great effort to gather them in until fall when they were either enticed or driven into pens for fattening on maize (corn). The corn-hog team goes back to the earliest times in this country. The Indians developed a taste for fat pork, but they made no effort to domesticate swine as they had domesticated horses and dogs. To them hogs were legitimate wild game, whether they were several generations wild or just the spring pig crop running in the woods.

SOON BECAME WILD

Also in the British colonies to the north, hogs arrived with the first settlers. While the breeding of these animals may have been somewhat different from the Spanish stock in the south, the early history of the swine from Britain was in many respects a repetition of the experi-

ence in the south. The northern pioneer farmers were somewhat more settled than the Spaniards; therefore fewer pigs ran wild. But the management was still pretty much the same. Let the pigs run wild until fall when they would be penned for fattening.

This early half-wild management scheme led to many other problems, such as rounding up the pigs for fattening, and controlling the number of boars which turned very quickly into bad actors. One writer recapturing frontier life tells of the farm boy whose job it was to castrate as many of the half-wild male pigs as possible. With some help from his dog he developed a technique. He would deliberately get himself chased by a herd of these clamoring pigs, a mixed bag of old boars, sows and small pigs. Climbing to safety in a previously selected tree he would skilfully drop a noose over a boar pig milling around just below, hoist the pig into the tree, castrate him with his sharp pocket knife, let him down, and then fish for another. Any farm boy who has held squirming pigs while they were being castrated by his father or older brother in the safety of a hog house will appreciate the dexterity of the boy in the tree.

RAISED FOR LARD

Another oddment of swine breeding in pioneer times which is hard to understand today is that hogs were raised for their lard, not for their meat. This remained true down through the 1800s which includes the period when most of the illustrations in this book were executed. One of the most profitable exports of animal products from the colonies to the home countries in Europe was lard, plus salt pork which was largely lard. Lard had many uses and was easier to keep than lean meat. In those days there was a great craving for fat in the human diet, probably because sources of vegetable fats had not been greatly developed. No doubt the strenuous physical activity of all humans of the time was a contributing factor.

The astonishing thing is that this craving for fat was more intense in the South than in the North. This persists even today when larger hogs and more lardy cuts of pork seem to find a better market in the South.

Here is an amazing paragraph, taken from Forbes' *History of Upper and Lower California,* which describes the nature of the hog business in that state during the Spanish period:

Hogs "are fed in a manner to produce as much fat and as little flesh as possible. They are allowed to grow to a certain stage in a lean state, subsisting chiefly on such roots and herbs as they can procure at large in the woods and fields, and when they arrive at the proper age and size for killing, they are shut up . . . and as much maize given them as they can eat. . . . By this means they get enormously fat, and when slaughtered they are found to be almost all lard. . . . This lard they peel off as blubber is peeled off a whale, the whole being entirely separated from the flesh and the entrails, leaving an astonishingly small proportion of flesh. . . . In the sale and purchase of these animals their weight of flesh is never taken into account; the calculation is how many pounds of lard they will produce."

Shades of the lean, long, lanky bacon hog which we strive for today!

ALSO GOOD TRAVELERS

The tales of hogs so lardy that they could hardly walk, and huge barrels of lard and salt pork filling ships on the way to Europe and to the rum ports in the Indies, somehow do not fit in well with other tales of the tough, leggy, razorback types that

101

IMPROVED CHESHIRE.

ABOVE

DUROC JERSEY hogs established themselves around 1900 as one of the most popular breeds in America. They are said to have acquired their red color from strains of Spanish swine from the south. They were first developed in New Jersey and spread over the country from that state.

AT LEFT

CHESHIRE, probably closely related to the Yorkshire, was a popular white hog for a time in the eastern states. This advertisement is from a Country Gentleman of 1884. Cheshire blood was probably transferred to the more popular modern breeds.

moved west with the covered wagons. One might even surmise that today's slim crossbreds and "hybrids" might make better travelers than the hogs of old. Not so!

The hog that won the West was inclined to be long of leg and snout, a tough, resourceful traveler, lean and fast in his working clothes. At the same time there existed an astonishing capability to settle down and "make a hog of himself" as a lard factory as soon as he was penned and confronted with unlimited amounts of corn. Historically minded animal nutritionists might find this capability an absorbing topic for research.

It is appropriate to quote here a paragraph from Towne and Wentworth describing the typical emigrant family unit on the trek West:

"Hazing along a few swine or sheep and a cow or two was his barefoot son on foot. On the wagon, drawn by a pair of plow horses or a yoke of oxen, was the grizzled pioneer himself, while inside were stowed his wife and younger chil-

AT RIGHT

UPPER ILLUSTRATION is of a pair of Poland Chinas, one of the most popular of the breeds in America around 1900. These were the property of D. M. Magie, Oxford, Ohio, illustrated in the Country Gentleman of 1876.

Middle illustration is of a group of Berkshire swine, owned by T. S. Cooper, Coopersburg, Pennsylvania, and illustrated in Country Gentleman in 1873. They are identified as Wharfdale Chief, Wharfdale Rose, and Hillhurst Rose. The artist no doubt took some liberties with the conformation of the hogs but did record the upturned nose which has recently become even shorter.

Lower illustration is of the Essex breed that was launched in this country as early as 1880 but evidently did not take. In Britain today the Essex is belted and looks very much like the American Hampshire.

dren, household goods, a coop full of chickens, some bacon and flour, and a cornhusk mattress. Slung from the axles under the wagon was often a wooden crate filled with eight or ten young pigs."

Yes, the pioneer hog was first and foremost a traveler. This capability must have persisted for a hundred years because the first half of the 1880s found hogs being driven on foot to market for hundreds of miles. Before Chicago became "hog butcher to the world" swine from Illinois and Indiana were moved in droves numbering in the tens of thousands, all the way to Cincinnati. Later Chicago became the principal hog market, and droves arrived regularly from Indiana, Illinois, Missouri, and Iowa. No doubt it was slow work, requiring months to negotiate the long drover trails, but hundreds of thousands of swine arrived at market this way— with enough fat left on them so that it was worth while to butcher them and ship enormous quantities of lard and barreled pork to the East.

AT LEFT

UPPER ILLUSTRATION is of a Yorkshire boar. This breed appeared on the American scene quite early as one of the popular white hogs. Only recently has it come into its own for its great contribution to crosses as a good milker and producer of large litters.

Middle illustration is of the Tamworth, another British breed that has played a very important part in the development of the bacon type hybrids of today. It is a red hog, an unusual color for Britain. Evidently it came to America early because it is here illustrated in an American magazine of the 1880s. The animal shown here is hardly of the bacon type, but like many other breeds this one has undergone a lot of change.

Lower illustration is of the Improved Lincoln, probably British in origin but not listed among the leading English breeds today. Are there any herds left in this country?

FIRST WEST, THEN EAST

As a matter of fact, the westward trek of settlers and breeding stock from eastern states was followed soon after by an eastward trek of hundreds of thousands of hogs being driven to markets in eastern cities. The nature and magnitude of this movement of hogs eastward boggles the mind of today's hog man who has his pigs ready for market in five months and sends them out to nearby markets in small batches the year round. How could so few hogs going west multiply into hordes going east in just a few decades?

The building of the railroads, with branch lines leading into practically every small town, advanced very rapidly in the last decades of the 1800s. This development turned the walking hog into a riding hog. The stage was set for more scientific hog breeding with the economic goals we are familiar with today: the "kind of cuts the consumer wants," the kind of hog that can be brought to market in six months or less, an average of ten pigs per litter, resistance to disease, and all the rest.

It should be borne in mind, however, that the demand for the lean bacon hog did not come soon. The lard hog remained very much in favor until well past the turn of the century. The breeders who undertook to tailor the hog for the market in the late 1880s had the lard hog in mind.

IMPROVED IN AMERICA

While the first hogs came from Europe, the favorite American breeds in the 1880s were very much an American project. There was some importing, but it did not attain the popularity and prestige that was true of horses and cattle. It was not until much later that the breeders went back to Europe in search of lean bacon types such as the Landrace and Tamworth. At the same time they went looking for genes that would ensure greater prolificacy (more pigs per litter) which seems to have gone into decline during the Golden Age of breed building in this country.

Cross breeding, which undoubtedly was rife in pioneer days and the woodsy era of hog raising, and is now standard practice in the most advanced "piggeries," was frowned upon in the breed building days. Out crossing was at first somewhat more tolerated than with other livestock because breeds were built pretty much from scratch.

The three most common "pure" breeds in America can be said to have originated in this country. They are Poland China, Chester White, and Duroc. Poland Chinas are believed to contain the nondescript blood of the hog most common during the trek west, plus infusions of a spotted hog known as the big China, an early breed known as Berkshire, the Irish Grazer, a sire brought from Poland, and perhaps others. By the 1860s the breed was identifiable and became known as Poland and China. Since then it has had a varied history, culminating in some very superior contemporary meat hogs that have repeatedly won carcass shows.

Chester (County) Whites evolved by careful breeding from a number of white strains, including Woburn or Bedford hogs of English origin, some white Chinese, and "Thin Rind" Norfolk stock, also from England; a White Normandy boar from France, and many other strains largely American.

When I was a small boy on the farm in 1912 a favorite cross was a Poland boar on a Chester White sow. At that time the Chesters had an excellent reputation as mothers. This is the cross that was most prominent in the late 1920s in launching the cross-breeding sanctioned by University animal husbandry specialists.

RED HOGS FROM JERSEY

Durocs originated in New Jersey from a strain of red hogs known as Jersey Reds. The red color is believed to have come from red Spanish hogs which were common in the South. The name Duroc came from a race horse, applied to this hog by a breeder who also worked at perfecting Jersey hogs. Red Berkshires from Connecticut and Red Rocks from Vermont may also have been merged into the Duroc breed.

There were many other breeds and strains, some of which have increased in popularity in very recent times. In the mid-1880s there came from England the Essex, Large and Small Yorkshires, and the Large Blacks. The latter had a part in building the Hampshire, a belted black and white which became very popular after the turn of the century. Cheshire was a white breed that started in New York State about 1855. This breed was almost indistinguishable from the White Yorkshire and was also related to the White Berkshire. A red and white breed called the Hereford, so named from markings similar to Hereford cattle, enjoyed limited popularity in the Midwest.

NEW SOURCES OF GENES

The rage for criss-crossing and "hybridization" of swine which has gone so far toward taking over the industry today had led to a search for genetic material loaded with characteristics considered most desirable in the modern consumer-tailored hog.

Fear is often expressed that we will run out of good material unless we keep up the pure breeds, which also are quite capable of being tailored to the modern market. The hog still has infinite possibilities for change and improvement. There is a lot of untried protoplasm of the family of *Sus scrofa* scattered over the world that can be brought into play by the hog breeders if they run out of good genes at home and in Europe.

AMERICAN AGRICULTURIST
FOR THE
⊹ FARM · GARDEN · & · HOUSEHOLD ⊹

"AGRICULTURE IS THE MOST HEALTHFUL, MOST USEFUL, AND MOST NOBLE EMPLOYMENT OF MAN."—WASHINGTON.

VOLUME L.—No. 6. NEW YORK, JUNE, 1891. NEW SERIES—No. 533.

THE MONARCH OF THE MOUNTAIN FLOCK.

SHEEP

Sheep were first domesticated, along with cattle and swine, shortly before the dawn of recorded history in the Neolithic period of man. This is the era that first saw agriculture practiced and is often called the first of the civilizations. The early sheep were more inclined to be hairy rather than wooly. There were at the time identifiable strains of sheep in Asia, Europe, Africa, and North America where the American Bighorn and its relatives have held sway without being successfully domesticated to this day.

The tame sheep in America in pioneer times, as with swine, goats, and cattle, were introduced by the Spaniards in the South and the English in the North. The sheep of the Navajo Indians in the Southwest are likely to be of Spanish origin. There may also be a good deal of this same blood in the breeds that we now call our Westerns.

It will surprise many to learn that the crossing of our European sheep with the Bighorn was attempted in this country in the latter quarter of the 1800s. Very recently some work along this line has been done by Ralph Yohe. However the cross-breeding of our truly native Bighorns with European sheep has never gained much headway.

Sheep are usually defined as hollow-horned ruminants, closely related to goats. Some of the earlier hairy sheep were, in fact, hard to distinguish from goats, but there are enough differences to discourage hybridization. Sheep and goats were considered mountain animals, and best suited for existence where vegetation is sparse. Sheep, however, have adapted themselves readily to rich lowland pastures and domestic feeding practices.

The European breeds of sheep appear to have been made from a wild European strain called the mouflon, plus an infusion of strains from Turkistan, Tibet, and other Himalayan areas. The Roman conquests and later the Crusades, which brought much intercourse between the East and the West, undoubtedly had much to do with the improvement of livestock.

IN BRITAIN BEFORE ROMANS

There were domesticated sheep in Britain even before the Romans arrived, but they probably brought along some sheep of their own to improve the native stock. History records that they established a factory for the production of wool cloth so they would not need to import clothing for their soldiers. Weaving with wool no doubt flourished in most parts of the civilized world from the time of Christ on. The earliest fibers to be used were flax and cotton. Flemish weavers of

AT LEFT

CAPTAIN JACK Jr., great Merino ram of American breeding, made a name for himself at the shows in the 1880s. While the Merino originated in Spain, superior strains were bred in the United States, Britain, Saxony, and other countries of the world. The Merino probably had wider influence in the improvement of sheep than any other breed.

wool were famous all through the Middle Ages, but it remained for the British to establish themselves as a kingdom built on wool, its production, trading, and weaving.

Since the climate and terrain of the British Isles was suited to sheep growing, and the development of the industrial revolution in England was founded largely on its textile mills, it was natural that the skill and ingenuity of English livestock breeders should center on breeding superior sheep as well as cattle. Nevertheless, the breed of sheep that has most influenced world development and trade was not British but Spanish.

This honor goes to the Spanish Merino. The Merino is a very hardy, fine-wooled sheep inclined to wrinkles in its hide and prominent horns. Improvement of the breed was undertaken largely outside of Spain. It received attention from the breeders in France where it had a famous descendant called the Rambouillet, in Saxony where it became the Saxon Merino and in England where it influenced many of the British and Scottish breeds. In America the Merino arrived as early as 1793, and was developed by breeders into a distinct strain called the American Merino. This was then amplified into several American strains. Later the Merino was to move to New Zealand and Australia where, after crossing with longwooled breeds, notably the Lincoln, it was line-bred into the New Zealand-Australian breed which is called the Corriedale.

Since Britain called the shots in the wool trade and was the principal market for world wool, Merino and British breed crosses took over in Canada and Argentina, as well as sections of Africa and Asia.

To begin with the Merino was distinguished for its hardiness in semi-arid sections of the world and for its fine wool. The mutton was of good quality but not especially abundant. The crosses with long-wooled (coarse-wooled) breeds produced a near ideal sheep for the range, with high yielding medium-wooled fleece.

ABOVE

ENTERPRISING BREEDERS made attempts to cross domestic European breeds with the American Bighorn as early as 1880. Above is an artist's conception of a Bighorn ram and his crossbred offspring. Bighorns do not take well to domesticity and are hard to handle in a breeding program.

One of the problems is building a fence high enough to hold them.

Lower illustration is of a Shropshire ram of the 1880s, shortly before the breeders became too much obsessed with wooled head and legs. For a long time the Shrop was the favorite breed on farms of the East and Midwest.

AMERICA ADOPTS BRITISH BREEDS

During the 1800s Americans became very much interested in the British breeds of which there were many. Importers got busy introducing and popularizing the different breeds. There were sheep on practically every farm, although in the East and Midwest they never became a major livestock enterprise. This occurred later as the West opened up to settlers and its agriculture assumed a more extensive aspect, namely the large ranch with additional grazing rights derived from the government.

The British breeds from which most American sheep are descended can be divided into three classes, the Longwools, the Shortwools, and the Mountain breeds.

Longwools. These get their name from their long wool which is also somewhat on the coarse side. The animals are mostly white-faced and hornless. Breeds usually placed in this class are Liecester, Border Liecester, Wensleydale, Cotswold, Lincoln, Devon Longwool, South Devon, and Romney Marsh.

Shortwools. These include the Down breeds, all of which are hornless with dark faces and legs, dense wool, and good mutton quality: Shropshire, Southdown, Hampshire Down, Dorset Down, Oxford Down, and Suffolk. Others are Dorset Horn, Western or Wiltshire, Ryeland, Kerry Hill, and Clun Forest.

Mountain. These are smaller, exceptionally hardy sheep which produce mutton of very high quality: Scotch Blackface, Lonk, Rough Fell, Cheviot, Welsh Mountain, Herdwick, and Exmoor.

In the making of the breeds in Britain, cross-breeding was widely practiced. Virtually all the named breeds are the result of crossing to get bigger and hardier animals, to alter the grade and yield of wool, to improve the quality of mutton, to enhance disease resistance and prolificacy. There followed a time when

UPPER ILLUSTRATION is of the Hampshire which began to gain in popularity in the United States in the early 1900s. Hampshires are a thrifty, growthy breed well suited for the small flock. They also do well in crosses.

The ewe in the lower illustration is a Border Leicester, close relative of the Leicester which probably contributed more to English sheep than any other breed. Leicesters, one of the oldest English breeds, were "longwools" that performed well in crosses with "shortwools" to give a hardy, medium wooled animal much in demand.

pure breeds were considered sacrosanct, especially in America where breed associations seem to have had more influence on the individual breeders. This was true especially in smaller purebred flocks which were shown at fairs and which furnished breeding rams for larger market flocks. In the range country out west cross-breeding was always in vogue. Flocks tended to turn up as Whiteface Westerns or Blackface Westerns, being a mixture of breeds that had proved themselves on the range and in the feedlot.

The decline of sheep growing in America has been cause for concern among livestock people who know sheep and realize that they can be a profitable enterprise. The rise of the synthetic fibers to replace cotton, linen, and wool has no doubt contributed to this decline. There is now a revival of interest in both cotton and wool. This trend stands a good chance of continuing, since petroleum from which most synthetics are made is bound to be scarcer and higher priced in the future. We hear more being said today about the qualities of the natural fibers that were somehow obscured during the stampede to the synthetics.

AT LEFT

ON THIS PAGE is a group of three English breeds: Upper illustration, a Horned Cheviot ram from a famous Scottish flock in 1889; middle, a first prize Dorset Horn ram of about the same time; lower, the Oxford Down ram Freeland, property of T. S. Cooper, Coopersburg, Pennsylvania, also in the 1880s. There is also a strain of Dorset sheep that is hornless.

CONSUMER ALLOWED
TO FORGET

Another thing that happened was that the American consumer got out of the habit of using lamb and mutton on the table. Such a trend tends to be cumulative, with disasterous results in the marketplace.

When the sheep raisers slugged it out with the cattlemen on the western range in the 1880s, both industries seemed to thrive. Although at no time were sheep considered unprofitable, good sheep herders became hard to find. Now that both the sheepherder and the cowboy have taken to the Jeep and the airplane it is even more difficult to find the skills and the patience to deal with sheep on a large scale.

On smaller farms in the East and Midwest, sheep somehow developed into a nuisance enterprise and suffered accordingly. The mechanization of farms, the disappearance of fences, and the advent of the corn-soybean-Miami rotation spelled doom for smaller auxiliary farm enterprises. Much feed is going to waste on our modern farms with their tendency toward confined livestock or no livestock at all.

AT RIGHT

ON THIS PAGE are three more English breeds: Upper illustration, an artist's conception of a purebred Cotswold ram owned by John Snell's Sons, Edmonton, Canada, in 1877, illustrated in the Country Gentleman of that year; middle, an English Southdown which became a popular breed in America; lower, Romney Marsh sheep which were very useful in England for grazing the marshes and were given some trial in this country without catching on. Hardiness was the strongest characteristic of this breed.

111

BREEDS IN TROUBLE

The pure breeds inherited the usual troubles that result from too close adherence to inbreeding and the styles of the show ring. We raised purebred Shropshires on the home farm in the period when it was mandatory for Shrops to have wooled faces and legs and small compact bodies. The heads became so wooled that the sheep could not see and consequently suffered in thrift. The wool on the legs contributed very little to the total wool weight. Finally the situation became so critical that one of the leading importers, Wm. McKerrow and Sons of Wisconsin, went back to England in the 1930s to get larger, bareface English Shropshires to restore the breed. Shropshires then emerged looking more like the Hampshires which had threatened to push them out of first place among American farm flocks.

Small breeding flocks on farms in the East and Midwest tended to be pampered and show ring oriented. They therefore suffered in hardiness and ability to produce practical market lambs. Large lamb feeders gravitated toward the western types.

The whims of the judges in the show rings have not always had a good effect on breeds of livestock. Like the creators of fashions in women's clothes, they had their likes and dislikes that were often aped by other judges. With fashions in livestock as with fashions in clothes, what the arbiters liked at a given time was not always the most practical.

The British have never ceased to breed good sheep. In recent decades they have broken away from the pure breeds to the extent that it is difficult to relate their sheep to pedigrees. The recent flurry of cross-breeding has turned up some excellent types which are still finding their way to other countries to improve the quality of sheep in general.

It should be mentioned here that losses from roaming dogs have been one of the principal reasons why the small farm flock was largely abandoned. The urbanization of the countryside in more heavily populated areas, with the old farmsteads and many new country homes occupied by commuters, has greatly increased the number of dogs. Many of these are permitted to run wild where they are joined by abandoned dogs loosed in the countryside by city people. These half-wild dogs have wrecked many a farm flock and discouraged many an otherwise ambitious young entrepreneur. Somehow this menace must be brought under control if sheep growing is to be revived.

BELOW

HERE IS A GROUP of Suffolk sheep from England. This breed never became widely known in this country but it has some very loyal adherents. Suffolks are known best for their fine quality of mutton. Some sheepmen like their clean faces.

EXCELLENT PROJECT
FOR YOUTH

When I was a boy, father regarded the flock of sheep as a project for the children of the family. He looked down his nose at sheep, but he did not disturb the enterprise because he recognized its value.

First the sheep supplied the spending money needs of us boys, then they took over the buying of our clothes, and lastly they paid for the education of one boy after another. I went through college on the proceeds from the sale of my share of the farm flock to my brothers.

Money was not the only value involved here. The keen interest which grew year by year, the independence of having our own source of funds, the knitting together of the interests of the children in the family—all paid rich dividends. It was a rule in the family that as soon as a boy reached six years of age, his older brothers were obligated to give him a ewe lamb of good breeding stock. The ewe lamb grew and multiplied. By the third year there was income which was the child's own to do with as he wished.

Some years later this plan was replaced on a wider scale by the 4-H project. This also proved to be a boon to farm boys and girls who often had a nice business started by the time they were through school and ready to start a career. If perchance the career was not farming, as in my case, the share of the business could be turned into a tidy sum to help out with the costs of higher education.

Farming is changing so rapidly that it is difficult to guess the pattern of the future. I have reason to believe that the part-time farmer of the future, who is replacing the small farmer of the past, will find small scale livestock growing to be both profitable and rewarding in other ways. If so I would recommend a flock of sheep as a very good enterprise for either the young or the old.

ABOVE

THESE ILLUSTRATIONS are very old and show the stylized approach of early livestock engravers. They set out to accentuate the features that breeders were working toward and ended up with what might be called a "charming distortion." The upper engraving is of a Leicester wether. It is from an issue of The Cultivator of the year 1845. The lower engraving is of the famous Lincoln ram Lord Chancellor, imported by Richard Gibson of Canada. It is from an early Country Gentleman.

113

OXFORDSHIRE DOWN EWE AND TWIN LAMBS—OWNED W. V. R. POWIS, WAYNE, ILL. 1886

AT RIGHT

UPPER ENGRAVING is that of the Herdwick, a small mountain breed little known outside of its native area in England. This ram, named Scawfell, got away from home long enough to win at British shows and get his picture in American farm magazines. The ram just below is a Scotch Blackface. This mountain breed was somewhat better known than the Lonk and the Herdwick. It was suited to high altitudes and sparse vegetation. The specimen shown here must have had a special combing job for show purposes, a luxury probably denied the rest of the flock. The mountain breeds were small sheep, wild and resourceful. Somehow they never took hold in America where there are also plenty of high altitude pastures.

LOWER ILLUSTRATION is of one of the many crosses between Merinos and native British breeds. This was called Thoroughbred Anglo-Merino, indicating that some line breeding had occurred to establish a breed.

OPPOSITE PAGE

THESE SHEEP are known as the Lonk, a very hardy mountain breed from the north of England. Not many were transplanted to the United States.

THE NEBRASKA FARMER.

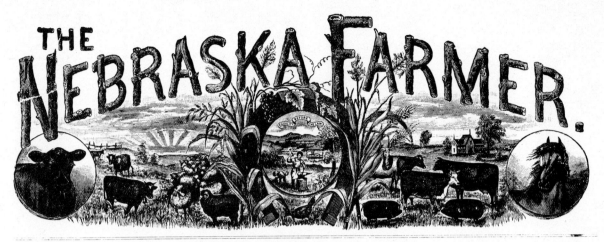

Vol. XX.
No. 14.

LINCOLN, NEB., THURSDAY, APRIL 2, 1896.

Whole No. 724.

FIRST PRIZE BREEDING PEN OF B. P. ROCKS AT MADISON SQUARE GARDEN, NEW YORK, OWNED BY A. C. HAWKINS, LANCASTER, MASS.

POULTRY

OPPOSITE PAGE

THE FRONT PAGE of the Nebraska Farmer serves a double purpose, to show the charming logos of the old farm papers, plus an excellent pen of Barred Plymouth Rocks, that were probably the favorite general purpose breed in the Midwest at the turn of the century. Rocks came in Buff, White and something called Partridge, all single-combed.

BELOW

EVERY POULTRY CATALOG of the late 1800s included some exotic breeds which could usually be dramatically illustrated in color. This pen of High-bred White Crested Polish fowls, appearing in a catalog, undoubtedly enticed many a farm boy or girl who begged the parents to get away from those dull gray Plymouth Rocks. Question: Were they any good at laying eggs or putting on meat?

Of the many enterprises that graced the old family farm and added salt and pepper to the lives of farm people, none has changed as much as the poultry business. When I was a boy a flock of chickens was basic to every farm operation. It was seldom a chief source of income. On most farms poultry were the responsibility of the womenfolk, aided and abetted by small boys who were not yet strong enough to wrestle teams of big horses or lift a man's load with pitchfork or scoop shovel.

Chickens gave the lady of the farmhouse her sole claim to financial independence. The chicken flock, often supplemented by geese, ducks, turkeys, guineas, and an occasional peafowl (the latter strictly ornamental), provided cash income from eggs, meat, and feathers. This income was controlled by the farmer's wife and was used for special things that made life more interesting for the family. From the chicken money came the means for the annual Christmas order sent off to Sears Roebuck or Montgomery Ward.

The chicken money was the chief contributor to the fund which brought a piano or an organ into the country home where it became an object of great admiration and a means of infusing culture into rustic urchins.

Nor should we neglect the value of the eggs and poultry meat as aids to the nutrition of the family. With all

117

that has been said and written about the bountiful meals served on the farm, we tend to forget that the cupboard in many farm homes was very bare indeed at certain times of the year. There was often little money with which to buy food at the store. "Company dinners" had to be put together from things available right on the farm. There was great need for small "luxury" supplements to plain fare which often consisted of bread, sorghum syrup, salt pork, and potatoes.

In an emergency the poultry flock came to the rescue. A favorite story in our family is about our Aunt Ida who raised a family out on the bare prairies of North Dakota. In dry years the larder was often practically empty. She would look down the road and see company coming two miles away (you could see a long way on those North Dakota plains). Aunt Ida was unflappable, as we say today. She would dash out into the poultry yard, chop the heads off a couple of hens, defeather and clean them, cut them up, and pop them into the big cast-iron frying pan. When the company drove into the farmyard in the slow lumber wagon or buckboard drawn by a team of work horses, the heavenly odor of frying chicken met them at the door.

AT LEFT

TOP: a very old engraving from The Cultivator of 1852, showing an exotic breed called Brahma Pootra. The birds were owned by Thomas Gould, Aurora, New York, in that year and shown at eastern fairs.

CENTER: These are imported Dark Brahmas, owned by T. S. Cooper, Coopersburg, New York, and shown in the Country Gentleman of 1873.

BOTTOM: These are rose-combed Silver Laced Wyandottes, a very popular general purpose breed. Wyandottes usually were white, but they were also available as Silver Laced and Golden Laced, combining beauty with utility.

118

THIS MAGNIFICENT ENGRAVING of mixed poultry was published in Prairie Farmer in 1886. The owner was F. W. Harding, Waukesha, Wisconsin. The artist had himself a ball putting grace and beauty into the birds. He wasn't far off the real thing at that. Prize chickens were beautiful, and the manners of the roosters squiring their hens around were impeccable.

BEFORE THE DAY
OF FREEZERS

That was hospitality at its noblest! Today's farm wife deals with unexpected company by going to the freezer from which she extracts various goodies that have been stored for the purpose. I doubt, however, that the supermarket broilers from the home freezer taste anything near as good as that fried chicken just an hour from the poultry yard.

It is hard to visualize the variety and extent of poultry breeding in the last half of the 1800s and the first quarter of the 1900s. There were hundreds of breeders, more often called fanciers. That is certainly what most of them were! They searched the world for exotic and highly ornamental breeds. They patronized the huge poultry shows. What has been called the first American show was held in Boston in 1848. There was much talk about egg laying and meat producing qualities, but delivery on the promises was poor. The great emphasis at the shows was on color, conformation, style—appearance if you please.

The greatest traffic until well into the 1900s was in the heavier breeds, usually called general purpose. These were mostly of Asiatic origin. They were "sitters," which meant that they would go broody in the spring and insist on hatching their own or alien eggs. The Mediterranean or egg-laying breeds, of which the most popular was the Leghorn family, arrived at their greatest popularity later after small incubators had been perfected. These were the "nonsitters" which meant that artificial means of incubating eggs had to be used.

The farm size incubator did not come into general use until around the turn of the century. Indeed, the artificial incubator had been invented way back in the time of the Egyptians and Romans, but they were cumbersome affairs fired by wood and tended full time by someone who lived in the incubator room with the eggs.

Management of the farm flock around the turn of the century was a far cry from today's large automated operations. The family was deeply involved. Meticulous, day-to-day care, even hour-to-hour care, was the watchword of success. The main flock, usually a hundred hens of a general purpose breed, was kept in a poultry house separate from the other livestock. This house needed a good deal of insulation because chickens did not thrive in bitter cold.

Roosters were withheld from the flock until toward spring when fertile eggs had to be provided for spring production of new chicks. Depending on breeding and care, winter egg production was usually sparse. Many a housewife with a nice flock of a hundred or more pullets was hard put to it to find enough eggs at Christmas time to do her baking.

R. AGRI.

120

LONG DAYS BROUGHT ACTION

But when the days grew longer in February, the hereditary juices began to flow, and the egg laying season quickly went into high gear. There was a saying that in early spring "even a feather duster would lay eggs."

Well before the start of the egg laying season we would send for five roosters (for each hundred hens), or we would shop around among neighbors who were reputed to be very good with chickens. We did not like to mix breeds because we believed mixing would eliminate the few good qualities that were already present. When the roosters arrived in the strange flock they would put on a heck of a battle among themselves to establish the proper pecking order. They had to be watched carefully to keep them from seriously maiming or killing each other in this battle for supremacy. Once the proper chain of authority had been established there was no more trouble, but it can be surmised that the head rooster sired

AT RIGHT

RHODE ISLAND RED [upper illustration] was a general purpose breed that grew in popularity after the turn of the century. They were productive in both meat and eggs and have been used rather widely in the building of broiler hybrids in recent years. There are also Rhode Island Whites.

The lower engraving is of the Silver Spangled Hamburg, resplendant in color and very popular among fanciers. After all, poultry were raised to be admired during the Golden Age.

OPPOSITE PAGE

WHITE LANGSHANS were an interesting breed that did not become very popular in this country. They were single-combed and characterized by feathers on the legs.

more than his share of the next generation of chickens.

Nobody, not even the horse, has a greater place in the legends and literature of the ages than the proud rooster. His "song" (crow), his cavalier manners among his ladies, and his magnificent courage have been praised by poets since the dawn of history. Socrates as he drank the fatal poison hemlock charged a friend to pay his debt of a cock to Aesculapius. Peter was reminded by a cock's crow of his cowardice when he denied Christ. Chaucer's Nun's Priest's Tale featured the adventures of a rooster and hen.

ARISTOTLE STUDIED CHICKENS

Aristotle, who has been called the first great scientist, studied in detail the anatomy and reproductive functions of the chicken and reported his findings. Cocks were used extensively in temple sacrifices before the advent of Christianity, and somehow got involved in all kinds of dreams, omens, and weird superstitions. The cock's crow, referred to as his "song" by early writers, was not only a means of telling time but of interpreting prophecies.

In my youth we did not regard the rooster as a very reliable timekeeper, but his crow was one of the familiar sounds of the countryside. Today, unfortunately, even in the country many a boy or girl will grow up without hearing this heavenly sound, so penetrating, persistent, and full of optimism.

In these days of large scale, automated poultry production the few roosters that are needed ply their trade in large breeding flocks, confined and virtually unknown. Egg production is in batteries of tens of thousands of hens. Chicken meat comes from broiler factories where production is the key word and there is no need for the song or the manners of the rooster.

This has led to the story of the bat-

ABOVE

LEGHORNS have always led the field among the Mediterranean breeds that were the backbone of the egg industry and the foundation stock of the many numbered hybrids that have taken over the egg field *today. In most parts of the country white eggs are most popular, accounting in part for Leghorn popularity. The pen of chickens shown just above is single-combed Brown Leghorn. The top illustration is of the Black Minorca which, together with the Black Ancona, is related to the Leghorn.*

tery bred hen that escaped from the laying house one day, and immediately exercised a hen's privilege of crossing the highway. She was run over by a Volkswagon. She got up, shook herself, and mused, "That must have been one of those rough roosters I keep hearing about!"

THE SETTING HEN

I must get back to the good old days when the rooster was the head man and cavalier around the farmyard. On our farm we were slow in utilizing the services of the mechanized incubator, so we had to rely on the old-fashioned setting hen. In middle spring after a hen had laid her quota of eggs she would turn broody. She found a nest which she liked and squatted on a clutch of eggs, settling down to twenty-one days of incubation. If the hen had hidden her nest in tall weeds or under a manger in the horse barn, the eggs might be her own. But she was not particular. Any eggs would do, even valuable eggs that might have been purchased from a poultry fancier and shipped hundreds of miles carefully packed in a basket of oats. Fifteen eggs were considered just right for a good sized setting hen.

It was my responsibility as a boy to look after the setting hens before I went to school in the morning. A completely free hen roaming the farmyard and choosing her own nest would take care of herself. But loose

AT RIGHT

THE GROTESQUE fowl shown in the top illustration is a Buff Cochin hen, described as an English breed produced entirely to put meat on the table and a gleam in the eye of the fancier.

The lower illustration is of the Red Cap, one of the handsome breeds that enjoyed a brief period of adulation at the poultry shows.

123

hens were considered notoriously unreliable. The spectre of management was present even in those simple times. The free setting hen might be driven off her nest by a rival, or worse yet, other hens might deposit fresh eggs in a partially incubated clutch. Waste was not to be tolerated if it could be avoided, for the egg money was a cornerstone of the farm family's well being.

PROJECT FOR A BOY

It fell to my lot to "manage" the setting hens in a fenced-in area of the hayloft which was by spring nearly empty. Setting hens were confined in rows of nests with slatted doors and released once a day for food, water, evacuation, and a bit of exercise. They were returned to their own nests. Hens might look alike to the

DUCKS, GEESE, AND TURKEYS were auxiliary poultry on the American farm. They may have deserved more attention than they got. Their eggs were not considered a commercial product but were used almost entirely for propagation. They were meat producers primarily, utilizing farm resources that were not used by chickens. Ducks and geese liked water, and the turkey, harking back to wild ancestors of the Pilgrim era, liked to roam.

The turkey illustrated here is a Bronze, undoubtedly most closely related to the wild turkey. There were also white turkeys which were used in breeding the modern meat bird of the supermarket. At top left is a pair of Toulouse geese, which became the most popular breed in America. The two most accepted duck breeds were the White Pekin and the Rouen. The latter looks a little like the wild Mallard. Both are illustrated here. Both contributed to the hybrid which furnishes ducklings for the market today.

Lower right is an interesting South American duck which enjoyed brief popularity. It was called the Muscovy.

novice or stranger, but the experienced poultry boy, like the shepherd, knew every individual even in a flock of a hundred.

The fuzzy chicks arrived promptly at 20 or 21 days. They were worth all the work and trouble. The surly, broody hen turned overnight into a busy, clucking, resourceful, and jealous mother. Given no help from humans, she would make out fairly well feeding the little fluffies with what could be gleaned in the farmyard. But of course we had to get into the act by feeding them bits of hardboiled eggs the first day and oatmeal or cracked grain from then on until they were large enough to range widely. Much could be written about the trials and tribulations of the hen and chicks, but there is not room here.

The chicks usually grew and prospered. They turned out to be about half cockerels and half pullets. The cockerels soon began to practice the art of crowing, while the pullets preened their feathers and looked coy. By fall they were roosting in the trees. With the arrival of the first cold nights in October, we "caught them in" and confined them to the laying house. As soon as possible we weeded out the cockerels to be shipped to market, and set the pullets up in clean laying quarters.

The mother hens reluctantly went back to their laying status, but their production wasn't impressive, especially during the late summer moult when they would quit entirely.

END OF THE LINE

As for the old roosters, their day of glory was over along about middle of June when we no longer wanted fertile eggs. Their careers usually ended on the Sunday dinner table, or as the main dish at the Fourth of July community picnic. They were a little on the tough side but their flavor was superb. A rooster that has chased grasshoppers and hens for a living, has supervised the affairs of the farmyard, and wakened the farmer's family every morning with his clarion call probably deserved better than to end up on the table. But that was his life's pattern. He was flavorful to the end—especially when compared with the lazy, pampered, juvenile poultry we buy in the supermarket today.

Early in the 1900s the imperfect system of perpetuating the flock by setting eggs under hens gave way to the mechanical incubator, a contrivance that could be heated by kerosene (later by electricity) and which could hold hundreds of eggs on trays in the warming chamber. The incubators had very indifferent controls at first and required almost as much tinkering as a Model T Ford, but they were an improvement on the setting hens. Their greatest advantage was that they could fill the gap for the non-setting breeds that were becoming more popular. The farm incubator gave way to the big commercial hatchery model as the 1900s progressed, which quickly opened the way for the factory-type poultry industry we have today.

I have included in this book illustrations of as many of the breeds as there was room for. The parade gives the reader an idea of the variety and excitement that was normal in the poultry business before it became homogenized in the interests of factory production of both eggs and poultry meat. I have tucked the poultry enterprise in with the livestock because it had an important place in the life and economy of the family farm.

NUMBERS INSTEAD OF BREEDS

The laying hens in the big modern commercial flocks are known by number rather than breed name. They are the product of meticulous cross-breeding and inbreeding, engineered for a single purpose, to produce lots of big eggs of uniform quality for one year, after which the hens go to the soup factory and are replaced by a younger set. The production of broilers is on the same scale and follows roughly the same pattern. The goal here is to get a very tender meat bird that reaches market in the fewest possible weeks with the least feed.

Feeding of both egg flocks and broiler flocks is computerized to give near perfect nutrition from the cheapest feed sources. This has given the consumer cheaper, standardized eggs and meat in the supermarket, but the bland flavor of the factory product leaves much to be desired. That is at least the opinion of many older consumers who once knew a different system and a different taste.

GOLDEN AGE OF POULTRY

The period celebrated by this book was indeed a Golden Age in poultry breeding, when poultry husbandry was fun and the presence of poultry running at large much of the year gave character and excitement to the farmyard. I have spoken repeatedly of personality in individual farm animals. I could say the same for chickens, or ducks, or geese! There were real characters in every flock. There was beauty too, for chickens came in gorgeous hues. They liked to preen their feathers and put on a fashion show, often to the detriment of the serious but drab business of laying eggs and putting on weight for market.

Ducks and geese have been somewhat neglected in America, even during the Golden Age. In Europe they get more attention. For instance, ducks can beat even chickens in rate of gain, and they have bred strains in Europe that will lay two eggs a day instead of just one. Geese have the advantage of being better grazers than any other kind of poultry. They have been pressed into service in some places to weed the grass out of strawberries without bothering the strawberry plants. Their sex life has caused some annoyance to breeders who expect one gander to take care of a dozen wives. The instinct in the goose family is for a gander to pick one goose and stick to her for life, eschewing the promiscuity which is the general rule among farm livestock. Cold-blooded breeders for profit do not approve of this kind of constancy. Both geese and ducks have been raised for their feathers as well as their meat, and to a lesser degree, eggs. They also like to have a lake or a stream nearby to play around in, although this is not an absolute necessity.

CLOSE COMPANION OF MAN

Wherever man dwells he is likely to be surrounded by poultry of various kinds that provide a part of his keep. He may even have a sporting interest in them. Fighting cocks are still in vogue in many parts of the world. Perhaps the craving for recreation may some day lead to a revival of interest in breeds of chickens, for beauty, and variety and adventure, rather than supermarket eggs and broiler meat.

Even the practical breeders may profit from such interest. After all, they made factory strains out of the old breeds. They may need some new genes again to develop disease resistance, better flavor, greater prolificacy, and other characteristics that make a better factory chicken. The only source of such genes will be old breeds preserved or exotic strains discovered in some far corner of the world where homogenization is still unknown.

Bibliographical Comment

This book is merely a glimpse into the past world of farm livestock which grows less familiar as more and more Americans grow away from the farm. This alienation extends even to active farmers because of the nature of modern agriculture. Many farmers today raise no livestock whatever. Others are exposed only to factory patterns of livestock production which lack what might be called the companionship factor that was so strong in the old days when farm people practically lived with their animals and knew them as individuals.

It is my hope that the glimpse afforded by this book will be an invitation to the reader to dig deeper into this fascinating relationship. Certainly it is an important part of our history, serving to humanize our past and give background to the economic and political forces which we think of as predominant in the building of America.

As stated in the Foreword, my purpose was partly to preserve and make more widely known the excellent engravings that are buried in the old files of farm magazines, now endangered by trends toward microfilming and efficiency in our libraries. The other purpose is to invite you to learn more about livestock, just for the fun of it and to help you understand your own past.

The following is not a formal bibliography but a few suggestions to help you in your search.

1. Never pass up a bound volume of an old farm magazine—anything dated before 1900. While once plentiful, these are now becoming very scarce. They may be hard to read but they are pure history in the making, almost as good as finding a bundle of letters written by your great great grandmother.

2. Old copies of the Department of Agriculture Yearbook are also great source material. Before the formal organization of the Department in 1862, an equivalent of the yearbook was published by the U. S. Patent Office which looked after the affairs of agriculture at the time.

3. Perhaps the best collection of modern books on the history of livestock has been published by University of Oklahoma Press at Norman, Oklahoma. It would pay to write them and ask for a complete list of their books in print. Three of their books I recommend highly as fascinating reading and good history, all by Charles Wayland Towne and Edward Norris Wentworth. They are *Shepherd's Empire* (1945), *PIGS: from Cave to Cornbelt* (1950), *Cattle and Men* (1955). This publisher has many more. I also recommend a most remarkable translation from the Latin called *Aldrovandi on Chickens,* edited and translated by L. R. Lind from manuscripts of around 1600. This is also an Oklahoma book. Other university presses have some excellent books that relate to the history of livestock. Send for lists.

4. For sheer pleasure in reading there is no book quite like *The Longhorns* by J. Frank Dobie. I believe this book was originally brought out in 1941 by Little, Brown and Company. My copy is a paperback issued by Grosset and Dunlap in the Universal Library series.

5. It is high time some of the old classics on livestock were reissued as paperbacks, because they have been out of print for many years. Here are a few of which I have copies but which are hard to find even in university libraries:

Henry S. Randall, *The Practical Shepherd,* 1864, D. D. T. Moore, Rochester, N.Y. Also by Randall, *Sheep Husbandry,* 1860, Orange Judd and Company, New York.

Henry S. Randall, *Youatt on the Structure and Diseases of the Horse,* 1852, C. M. Saxon, New York. Youatt was an English authority who is credited with many writings on the history of livestock. He is respectfully referred to as Mr. Youatt (no first name or initials) in very old American books.

Richard L. Allen, *Domestic Animals, History and Description,* 1847, Orange Judd and Company, New York.

Lewis F. Allen, *American Cattle, History of the Short-Horn Cattle,* 1874, and other books on rural life.

Alvin H. Sanders, *At the Sign of the Stock Yard Inn,* 1915, Sanders Publishing Company.

The above suggestions are just "teasers." History of livestock is in fact a much neglected topic and might well deserve more serious attention of research people.

The Author

Paul C. Johnson, the author of *Farm Animals,* has spent much of his life as an agricultural editor and student of agricultural history. He was editor of *Prairie Farmer* and editorial director of the *Farm Progress Publications* for twenty-two years until his retirement in 1970. Before that he was a country newspaper editor for ten years and a member of the faculty of the University of Minnesota Institute of Agriculture for seven years. During his years as editor, Johnson traveled widely and collected books and anecdotes on American farm history. His collection is probably the most extensive in this country. He still writes the John Turnipseed column in *Prairie Farmer* and has produced a book of Turnipseed pieces called *John Turnipseed's Four Seasons Almanac,* published by Wallace-Homestead Book Company.

Upon retirement, Johnson moved back to Northfield, Minnesota, which is near the home farm where he was reared. He and his wife Eveline, who has been his partner in all his publishing adventures including this one, are very busy editing books on antiques, traveling to visit antiques shows and museums, and "boning up" on rural history.